Lecture Notes
in Control and Information Sciences   254

Editor: M. Thoma

Springer
London
Berlin
Heidelberg
New York
Barcelona
Hong Kong
Milan
Paris
Singapore
Tokyo

Barbara Hammer

Learning with Recurrent Neural Networks

With 24 Figures

Springer

Series Advisory Board

Author

Barbara Hammer, Dr rer. nat., DiplMath
Department of Mathematics and Computer Science, University of Osnabrück ,
D-49069 Osnabrück, Germany

ISBN 1-85233-343-X Springer-Verlag London Berlin Heidelberg

British Library Cataloguing in Publication Data
Hammer, Barbara
    Learning with recurrent neural networks. - (Lecture notes
    in control and information sciences)
    1.Neural networks (Computer science) 2.Machine learning
    I.Title
    006.3'2
    ISBN 185233343X

Library of Congress Cataloging-in-Publication Data
A catalog record for this book is available from the Library of Congress

Typesetting: Camera ready by author
Printed and bound at the Athenæum Press Ltd., Gateshead, Tyne & Wear
69/3830-543210 Printed on acid-free paper SPIN 10771572

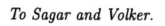

To Sagar and Volker.

Preface

A challenging question in machine learning is the following task: Is it possible to combine symbolic and connectionistic systems in some mechanism such that it contains the benefits of both approaches? A satisfying answer to this question does not exist up to now. However, approaches which tackle small parts of the problem exist. This monograph constitutes another piece in the puzzle which eventually may become a running system. It investigates so-called folding networks – neural networks dealing with structured, *i.e.*, symbolic inputs.

Classifications of symbolic data in a connectionistic way can be learned with folding neural networks which are the subject of this monograph. This capability is obtained in a very natural way via enlarging standard neural networks with appropriate recurrent connections. Several successful applications in classical symbolic areas exist – some of which are presented in the second chapter of this monograph. However, the main aim of this monograph is a precise mathematical foundation of the ability to learn with folding networks. Apart from the in-principle guarantee that folding networks can succeed in concrete tasks, this investigation yields practical consequences: Bounds on the number of neurons which are sufficient to represent the training data are obtained. Furthermore, explicit bounds on the generalization error in a concrete learning scenario can be derived. Finally, the complexity of training is investigated.

Moreover, several results of general interest in the context of neural networks and learning theory are included in this monograph since they form the basis of the results for folding networks: Approximation results for discrete time recurrent neural networks, in particular explicit bounds on the number of neurons and a short proof of the super-Turing universality of sigmoidal recurrent networks, are presented. Several contributions to distribution-dependent learnability, an answer to an open question posed by Vidyasagar, and a generalization of the luckiness framework are included. The complexity of standard feed-forward networks is investigated and several new results on the so-called loading problem are derived in this context.

Large parts of the research reported in this monograph were performed while I was preparing my Ph.D. thesis in Theoretical Computer Science at the Universiy of Osnabrück. I am pleased to acknowledge Thomas Elsken,

Johann Hurink, Andreas Küchler, Michael Schmitt, Jochen Steil, and Matei Toma for valuable scientific discussions on the topics presented in this volume. Besides, I had the opportunity to present and improve parts of the results at several institutes and universities, in particular during visits at Rutgers University (U.S.A.) and the Center for Artificial Intelligence and Robotics (Bangalore, India). I am deeply indebted to Bhaskar Dasgupta, Eduardo Sontag, and Mathukumalli Vidyasagar for their hospitality during my stay at Rutgers University and CAIR, respectively. Furthermore, I gratefully acknowledge the people whose encouragement was of crucial support, in particular Grigoris Antoniou, Ute Matecki, and, of course, Manfred. Special thanks go to Teresa Gehrs who improved the English grammar and spelling at the same rate as I introduced new errors, to Prof. Manfred Thoma as the Editor of this series, and Hannah Ransley as the engineering editorial assistant at Springer-Verlag. Finally, I would like to express my gratitude to my supervisor, Volker Sperschneider, who introduced me to the field of Theoretical Computer Science. He and Mathukumalli Vidyasagar, who introduced me to the field of learning theory – first through his wonderful textbook and subsequently in person – laid the foundation for the research published in this volume. I would like to dedicate this monograph to them.

Osnabrück, April 2000.

Table of Contents

Chapter 1

Introduction

In many areas of application, computers far outperform human beings as regards both speed and accuracy. In the field of science this includes numerical computations or computer algebra, for example, and in industrial applications hardly a process runs without an automatic control of the machines involved. However, some tasks exist which turn out to be extremely difficult for computers, whereas a human being can manage them easily: comprehension of a spoken language, recognition of people or objects from their image, driving cars, playing football, answering incomplete or fuzzy questions, to mention just a few. The factor linking these tasks is that there is no exact description of how humans solve the problems. In contrast to the area in which computers are used very successfully, here a detailed and deterministic program for the solution does not exist. Humans apply several rules which are fuzzy, incomplete, or even partially contradictory. Apart from this explicit knowledge, they use implicit knowledge which they have learned from a process of trial and error or from a number of examples. The lack of exact rules makes it difficult to tell a robot or a machine to behave in the same way. Exact mathematical modeling of the situation seems impossible or at least very difficult and time-consuming. The available rules are incomplete and not useful if they are not accompanied by concrete examples.

Such a situation is addressed in the field of machine learning. Analogously to the way in which humans solve these tasks, machines should be able to learn a desired behavior if they are aware of incomplete information and a number of examples instead of an exact model of the situation. Several methods exist to enable learning to take place: Frequently, the learning problem can be formulated as the problem of estimating an unknown mapping f if several examples $(x \mapsto f(x))$ for this mapping are given. If the input and output values are discrete, $e.g.$, symbolic values, a symbolic approach like a decision tree can be applied. Here the input for f consists of several features which are considered in a certain order depending on the actual input. The exact number and order of the considered attributes is stored in a tree structure, the actual output of a special input can be found at the leaves of such a decision tree. Learning means finding a decision tree which represents a mapping such that the examples are mapped correctly and the entire tree has the most simple structure. The learning process is successful if not only the known examples but even the underlying regularity is represented by the decision tree.

In a stochastic learning method, the function class which is represented by the class of decision trees in the previous learning method is substituted by functions which correspond to a probability distribution on the set of possible events. Here the output in a certain situation may be the value such that the probability of this output, given the actual input, is maximized. Learning means estimating an appropriate probability which mirrors the training data.

Neural networks constitute another particularly successful approach. Here the function class is formed by so-called neural networks which are motivated by biological neural networks, that means the human brain. They compute complex mappings within a network of single elements, the neurons, which each implement only very simple functions. Different global behavior results from a variation of the network structure, which usually forms an arbitrary acyclic graph, and a variation of the single parameters which describe the exact behavior of the single neurons. Since the function class of neural networks is parameterized, a learning algorithm is a method which chooses an appropriate number of parameters and appropriate values for these parameters such that the function specified in this way fits to the given examples.

Artificial neural networks mirror several aspects of biological networks like a massive parallelism improving the computation speed, redundancy in the representation in order to allow fault tolerance, or the implementation of a complex behavior by means of a network of simple elements, but they remain a long way away from the power of biological networks. Artificial neural networks are very successful if they can classify vectors of real numbers which represent appropriately preprocessed data. In image processing the data may consist of feature vectors which are obtained via Fourier transformation, for example. But when confronted with the raw image data instead, neural networks have little chance of success. (From a theoretical point of view they should even succeed with such complex data assuming that enough training examples are available – but it is unlikely that success will come about in practice in the near future.) The necessity of intense preprocessing so that networks can solve the learning task in an efficient way is of course a drawback of neural networks. Preprocessing requires specific knowledge about the area of application. Additionally, it must be fitted to neural networks, which makes a time-consuming trial and error process necessary in general. But this problem is not limited to the field of neural networks: Any other learning method also requires preprocessing and adaption of the raw data. Up to now, no single method is known which can solve an entire complex problem automatically instead of only a small part of the problem.

This difficulty suggests the use of several approaches simultaneously, each solving just one simple part of the learning problem. If different methods are used for the different tasks because various approaches seem suitable for the different subproblems, a unified interface between the single methods is necessary such that they can exchange their respective data structures. Additionally, the single methods should be able to deal with data structures which are suitable to represent the entire learning methods involved. Then they can control parts of the whole learning process and partially automate the modularization of the original task.

Here a problem occurs if neural networks are to be combined with symbolic approaches. Standard neural networks deal with real vectors of a fixed dimension, whereas symbolic methods work with symbolic data, *e.g.*, terms, formulas, ..., *i.e.*, data structures of an in principle unbounded length. If standard networks process these data, either because they classify the output of another learning method or because they act as global control structures of the learning algorithms involved, it becomes necessary to encode the structured data in a real vector. But universal encoding is not generally fitted to the specific learning task and to the structure of the artificial network which uses the encoded values as inputs. In an encoding process an important structure may be hidden in such a way that the network can hardly use the information; on the contrary, redundant encoding can waste space with superfluous information and slow down the training process and generalization ability of the network.

An alternative to encoding structured data is to alter the network architecture. Recurrent connections can be introduced such that the network itself selects the useful part of the structured data. Indeed, recurrent networks are able to work directly on data with a simple structure: lists of *a priori* unlimited length. Therefore, encoding such data into a vector of fixed dimension becomes superfluous if recurrent networks are used instead of standard feedforward neural networks. A generalization of the recurrence to arbitrary tree structures leads to so-called folding networks. These networks can be used directly in connection with symbolic learning methods, and are therefore a natural tool in any scenario where both symbolic data and real vectors occur. They form a very promising approach concerning the integration of symbolic and subsymbolic learning methods.

In the following, we will consider recurrent and folding networks in more detail. Despite successful practical applications of neural networks – which also exist for recurrent and folding networks – the theoretical proof of their ability to learn has contributed to the fact that neural networks are accepted as standard learning tools. This theoretical investigation implies two tasks: a mathematical formalization of learnability and a proof that standard neural networks fit into this definition. Learnability can be formalized by Valiant's paradigm of 'probably approximately correct' or PAC learnability. The possibility of learning with standard neural networks in this formalism is based on three properties: They are universal approximators, which means that they are capable of approximating anything that they are intended to approximate; they can generalize from the training examples to unseen data such that a trained network mirrors the underlying regularity that has to be learned, too; and fitting the network to the training examples is possible with an efficient training algorithm.

In this monograph, we will examine the question of approximation capability, learnability, and complexity for folding networks. A positive answer to these questions is a necessary condition for the practical use of this learning approach. In fact, these questions have not yet been entirely solved for recurrent or standard feed-forward networks and even the term of 'probably approximately correct' learnability leaves some questions concerning the formalization of learnability unsolved. Apart from the results which show the theoretical possibility of learning with folding networks we will obtain results which are of interest in the field of recurrent networks, standard feed-forward networks, or learnability in principle, too.

The volume consists of 5 chapters which are as independent as possible of each other. Chapter 2 introduces the general notation and the formal definition of folding networks. A brief description of their in-principle use and some concrete applications are mentioned. Chapters 3 to 5 each examine one of the above mentioned topics: the approximation ability, learnability, and complexity of training, respectively. They each start with a specification of the problems that will be considered in the respective chapter and some bibliographical references and end with a summary of the respective chapter. The monograph is concluded with a discussion of several open questions.

Chapter 2

Recurrent and Folding Networks

This chapter contains the formal definition of folding networks, the standard training algorithm, and several applications. Folding networks constitute a general learning mechanism which enables us to learn a function from a set of labeled k-trees with labels in some finite dimensional vector space into a finite dimensional vector space. By labeled k-trees we address trees where the single nodes are equipped with labels in some alphabet and every node has at most k successors. This sort of data occurs naturally in symbolic areas, for example. Here the objects one is dealing with are logical formulas or terms, the latter consisting of variables, constants, and function symbols, the former being representable by terms over some enlarged signature with function symbols corresponding to the logical symbols. Terms have a natural representation as k-trees, k being the maximum arity of the function symbols which occur: The single symbols are enumerated and the respective values can be found in the labels of a tree. The tree structure mirrors the structure of a term such that subtrees of a node correspond to subterms of the function symbol which the respective node represents.

Hence folding networks enable us to use connectionistic methods in classical symbolic areas. Other areas of application are the classification of chemical data, graphical objects, or web data, to name just a few. We will show how these data can be encoded by tree structures in the following. But first we formally define folding networks which constitute a generalization of standard feed-forward and recurrent networks.

2.1 Definitions

One characteristic of a neural network is that a complex mapping is implemented via a network of single elements, the neurons, which in each case implement a relatively simple function.

Definition 2.1.1. *A* feed-forward neural network *consists of a tuple* $F = (N, \rightarrow, \mathbf{w}, \boldsymbol{\theta}, \mathbf{f}, I, O)$. N *is a finite set, the set of* neurons *or* units, *and* $(N, \rightarrow)$ *is an acyclic graph with vertices in* $N \times N$. *We write* $i \rightarrow j$ *if neuron* i *is connected to neuron* j *in this graph. Each connection* $i \rightarrow j$ *is equipped with*

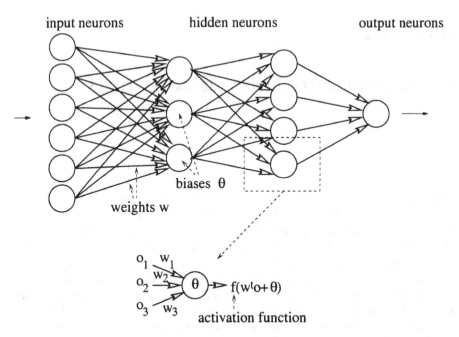

Fig. 2.1. Example for a feed-forward network with a multilayered structure; given an input, the neurons compute successively their respective activation.

a weight $w_{ij} \in \mathbb{R}$, **w** *is the vector of all weights. The neurons without pre-decessor are called* input neurons *and constitute the set* I. *All other neurons are called* computation units. *A nonempty subset of the computation units is specified and called* output units, *denoted by* O. *All computation units, which are not output neurons, are called* hidden neurons. *Each computation unit* i *is equipped with a bias* $\theta_i \in \mathbb{R}$ *and an* activation function $f_i : \mathbb{R} \to \mathbb{R}$. $\boldsymbol{\theta}$ *and* **f** *are the corresponding vectors. We assume w.l.o.g. that* $N \subset \mathbb{N}$ *and that the input neurons are* $\{1, \ldots, m\}$.

A network with m *inputs and* n *outputs computes the function*

$$f : \mathbb{R}^m \to \mathbb{R}^n \ , \ f(x_1, \ldots, x_m) = (o_{i_1}, \ldots, o_{i_n})$$

where $i_1, \ldots, i_n$ *are the output units and* o_i *is defined recursively for any neuron* i *by*

$$o_i = \begin{cases} x_i & \text{if } i \text{ is an input unit,} \\ f_i(\sum_{j \to i} w_{ji} o_j + \theta_i) & \text{otherwise.} \end{cases}$$

The term $\sum_{j \to i} w_{ji} o_j + \theta_i$ *is called the* activation *of neuron* i.

An architecture *is a tuple* $\mathcal{F} = (N, \to, \mathbf{w}', \boldsymbol{\theta}', \mathbf{f}, I, O)$ *as above, with the difference that we require* **w**' *and* $\boldsymbol{\theta}'$ *to specify a weight* w'_{ij} *for only some of the connections* $i \to j$, *or a bias* θ'_i *for some of the neurons* i, *respectively.*

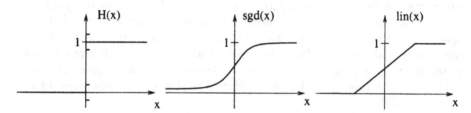

Fig. 2.2. Several activation functions: perceptron activation H, standard sigmoidal function sgd, and semilinear activation lin; all three functions are squashing functions.

In an obvious way, an architecture stands for the set of networks that results from the above tuple by completing the specification of the weight vector and biases. By an abuse of notation, the set of functions that result from these networks is sometimes called an architecture, too. Since these functions all have the same form, we sometimes denote the architecture by a symbol which typically denotes such a function. For a concrete network F the *corresponding architecture* refers to the architecture where no w_{ij} or θ_i are specified. A network computes a mapping which is composed of several simple functions computed by the single neurons. See Fig.2.1 as an example. The activation functions f_i of the single computation units are often identical for all units or all but the output units, respectively. We drop the subscript i in these cases. The following activation functions will be considered (see Fig.2.2):

The *identity* id $: \mathbb{R} \to \mathbb{R}$, $\mathrm{id}(x) = x$, which is often used as an output activation function in order to implement a scaling or shifting of the desired output domain; the *perceptron activation* H $: \mathbb{R} \to \mathbb{R}$,

$$H(x) = \begin{cases} 0 & x < 0 \\ 1 & x \geq 0 \end{cases},$$

which is a binary valued function occurring mostly in theoretical investigations of networks. In practical applications the activation functions are similar to the perceptron activation for large or small values and have a smooth transition between the two binary values. The *standard sigmoidal function*

$$\mathrm{sgd}(x) = (1 + e^{-x})^{-1}$$

and the scaled version $\tanh(x) = 2 \cdot \mathrm{sgd}(2x) - 1$ are common functions of this type. A more precise approximation of these sigmoidal activations which is sometimes considered in theoretical investigations is the *semilinear function*

$$\mathrm{lin}(x) = \begin{cases} 1 & x \geq 1 \\ x & x \in \,]0, 1[\\ 0 & x \leq 0 \end{cases},$$

which fits the asymptotic behavior of sgd and the linearity at the point 0. For technical reasons we will consider the *square activation* $x \mapsto x^2$. By a *squashing activation* we mean any function which is monotonous with function values that tend to 1 for large inputs x and to 0 for small input values x, respectively. A function is C^n if it is n times continuously differentiable. A property of a function holds *locally* if it is valid in the neighborhood of at least one point.

For technical reasons we will consider *multiplying units*, which simply compute a product of the output values of their predecessing units instead of a weighted sum, *i.e.*, $o_i = \prod_{j \to i} o_j$ for multiplying units i.

The connection structure $\to$ often has a special form: the neurons decompose into several groups $N_0, \ldots, N_{h+1}$, where N_0 are the input neurons, N_{h+1} are the output neurons, and $i \to j$ if and only if $i \in N_k$, $j \in N_{k+1}$ for some k. Such a network is called a *multilayer feed-forward network*, or *MLP* for short, with h *hidden layers*; the neurons in N_i constitute the hidden layer number i for $i \in \{1, \ldots, h\}$.

Feed-forward networks can handle real vectors of a fixed dimension. More complex objects are trees with labels in a real vector space. We consider trees where any node has a fixed fan-out k, which means that any nonempty node has at most k successors, some of which may be the empty tree. Hence, a *tree* with labels in a set Σ is either the empty tree which we denote by $\perp$ or it consists of a root which is labeled with some value $a \in \Sigma$ and k subtrees $t_1, \ldots, t_k$ some of which may be empty. In the latter case we denote the tree by $a(t_1, \ldots, t_k)$. The set of trees which can be defined as above is denoted by Σ_k^*. In the following, Σ is a finite set or a real vector space.

One can use the recursive nature of trees to construct an induced mapping which deals with trees as inputs from any vector valued mapping with appropriate arity:

Definition 2.1.2. *Assume R is a set. Any mapping $g : \Sigma \times R^k \to R$ and initial context $y \in R$ induces a mapping $\tilde{g}_y : \Sigma_k^* \to R$, which is defined recursively as follows:*

$$\begin{aligned} \tilde{g}_y(\perp) &= y, \\ \tilde{g}_y(a(t_1, \ldots, t_k)) &= g(a, \tilde{g}_y(t_1), \ldots, \tilde{g}_y(t_k)). \end{aligned}$$

This definition can be used to formally define recurrent and folding networks:

Definition 2.1.3. *A folding network consists of two feed-forward networks which compute $g : \mathbb{R}^{m+k \cdot l} \to \mathbb{R}^l$ and $h : \mathbb{R}^l \to \mathbb{R}^n$, respectively, and an initial context $\mathbf{y} \in \mathbb{R}^l$. It computes the mapping*

$$h \circ \tilde{g}_{\mathbf{y}} : (\mathbb{R}^m)_k^* \to \mathbb{R}^n.$$

A folding architecture is given by two feed-forward architectures $\mathcal{F}$ with $m + k \cdot l$ inputs and l outputs, and $\mathcal{G}$ with l inputs and n outputs and an only partially defined initial context $\mathbf{y}'$ in $\mathbb{R}^l$.

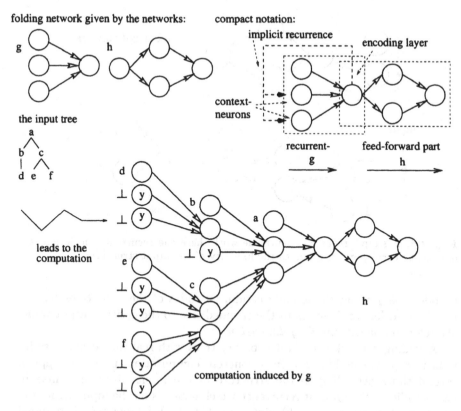

Fig. 2.3. Example for the unfolding process if a concrete value is computed with a very simple folding network; the folding network is formally unfolded according to the structure of the input tree; for every subtree one copy of the recursive part of the folding network can be found in the computation.

The input neurons $m + 1, \ldots, m + k \cdot l$ of g are called context neurons. g is referred to as the recursive part of the network, h is the feed-forward part. The input neurons of a folding network or architecture are the neurons $1, \ldots, m$ of g, the output units are the output neurons of h. If $k = 1$ we call folding networks recurrent networks.

Specifying all weights and biases in an architecture given by $\mathcal{F}$ and $\mathcal{G}$ and all coefficients in $\mathbf{y}'$ leads to a class of folding networks. As before, we sometimes identify the function class which can be computed by these folding networks with the folding architecture.

To understand how a folding network computes a function value one can think of the recursive part as an encoding part: a tree is encoded recursively into a real vector in $\mathbb{R}^l$. Starting at the empty tree $\bot$, which is encoded by the initial context $\mathbf{y}$, a leaf $a(\bot, \ldots, \bot)$ is encoded via f as $f(a, \mathbf{y}, \ldots, \mathbf{y})$ using the code of $\bot$. Proceeding in the same way, a subtree $a(t_1, \ldots, t_k)$ is

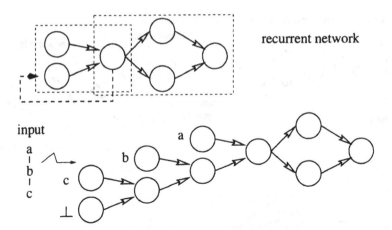

recurrent network

Fig. 2.4. A simple recurrent neural network: here the recurrent connection is a one-to-one correspondence due to the linearity of the input structure.

encoded via f using the already computed codes of the k subtrees $t_1, \ldots, t_k$. The feed-forward part maps the encoded tree to the desired output value. We refer to l as the *encoding dimension*.

A folding network which takes binary trees with real valued nodes as inputs is depicted in Fig. 2.3. In a concrete computation, part g is applied several times according to the structure of the input. Recurrent links in part g indicate the implicit recurrence which depends on the input structure. The input tree $a(b(d(\perp, \perp), \perp), c(e(\perp, \perp), f(\perp, \perp)))$, for example, is mapped to the value $h(g(a, g(b, g(d, y, y), y), g(c, g(e, y, y), g(f, y, y))))$. Note that for each fixed structure of input trees one can find an equivalent feed-forward network which computes the same output by simply unfolding the folding network. According to the recurrence, several weights are identical in this unfolded network.

For $k = 1$, the inputs consist of sequences of real vectors in $\mathbb{R}^m$. Here the recurrent connections have a one-to-one correspondence since the unfolding process is linear (see Fig. 2.4). In the case $k = 1$ we will drop the subscript k. Then a tree $(a_1(a_2(a_3(\ldots (a_n(\perp))\ldots))))$ is denoted by $[a_n, \ldots, a_2, a_1]$.

Any mappings we consider are total mappings unless stated otherwise. A mapping between real vector spaces is measurable or continuous, respectively, if it is measurable with respect to the standard Borel σ-algebra or continuous with respect to the standard topology, respectively. A mapping $f : (\mathbb{R}^m)^*_k \to \mathbb{R}^n$ is measurable or continuous if and only if any restriction of f to trees of a fixed structure is measurable or continuous as a mapping between real vector spaces. This latter requirement defines the topology and σ-algebra, respectively, on the set of trees. In particular, the set which consists of all trees of one fixed structure is open and measurable. If we restrict our consideration to trees of a fixed structure, a set is open or measurable if and only if it is open or measurable, respectively, with respect to the standard topology or σ-algebra on a real vector space. In the following, we denote by $\Sigma_k^{\leq t}$ the k-trees with height at most t and by Σ_k^t the k-trees with exactly height t and labels in Σ.

2.2 Training

In a concrete learning task several examples $(x_i, f(x_i))_{i=1}^p$, the *training set*, of an unknown function are available. Here the $x_i \in (\mathbb{R}^m)_k^*$ are vectors, sequences, or trees, depending on the learning task, and $f : (\mathbb{R}^m)_k^* \to \mathbb{R}^n$ is a vector valued function which is to be learned. For a fixed folding architecture $h \circ \tilde{g}_y$ with m inputs and n outputs, where all weights and biases are unspecified, we can define the *quadratic error*

$$E(\mathbf{w}) = \sum_{i=1}^p |f(x_i) - h \circ \tilde{g}_y(x_i)|^2$$

which depends on the weights and biases $\mathbf{w}$. Often, the initial context $\mathbf{y}$ is chosen as some fixed vector, the origin, for example. Minimizing $E(\mathbf{w})$ with respect to the weights yields a folding network which outputs on data x_i some value as similar as possible to the desired output $f(x_i)$. Since the activation functions used in practice are usually differentiable, the error minimization can be performed with a simple gradient descent method, *i.e.*, some procedure of the form:

```
w := some initial values
while E(w) is large
    w := w − η · ∇E(w)
```

with some value $0 < \eta$. For appropriate η and appropriate starting point $\mathbf{w}$, this converges to a local optimum of the function E with positive definite Hessian. The gradient $\nabla E(\mathbf{w})$ can be computed very efficiently, *i.e.*, in linear time with respect to the number of weights, using a method called *back-propagation* [105, 136, 138] for feed-forward architectures, *back-propagation through time* for recurrent architectures [90, 106, 137], and *back-propagation through structure* for folding architectures [41, 42, 73]. In order to speed up the above procedure one can adapt factor η appropriately or take second-order information into account. To avoid local minima one can use heuristic modifications of the above rule or add some form of randomization to the process [53, 103, 141].

Hence this algorithm allows us to find weights for a specific architecture such that the examples are fitted. Commonly, the architecture is chosen by a trial and error process. More precisely: One fixes several different architectures which seem appropriate – the proceeding chapters will give us hints of how this can be made more precise. Roughly, they yield upper bounds on the number of neurons such that the capability of bringing the quadratic error to zero or the in-principle possibility of valid generalization to unseen examples, respectively, is guaranteed. Hence the search space for an appropriate architecture becomes finite. After choosing several architectures, the quality of each architecture is estimated as follows: Each architecture is trained on one part of the training set with (a modification of) the above training method. Afterwards, the quadratic error on the remaining part of the training set is used to estimate the quality of the architecture. Obviously, one cannot use the quadratic error on the training set because in general, it will be smaller than the value one is interested in, the deviation of the network function from the function that is to be learned. Hence one uses the quadratic error on a set not used for training as an estimation of $\int |f(x) - h \circ \tilde{g}_y(x)|^2 dP$, P denoting the probability measure on the input trees in accordance to which the input patterns are chosen. Unfortunately, this method yields a large variance. The variance can be reduced if the single architectures are trained and tested several times on different divisions of the training data using so-called cross-validation [125]. Hence we get a ranking of the different architectures and can use the best architecture for the final network training as described above.

This roughly describes the in-principle learning method. Note that this yields a training method, but up to now no theoretical justification of its quality. For feed-forward networks the training method is well founded because their universal approximation capability and information theoretical and complexity theoretical learnability are (with some limitations) well understood. These theoretical properties do not transfer to folding networks immediately. The necessary investigations are the topic of this volume and altogether establish the in-principle possibility of learning with the method as described above.

However, a couple of refinements and modifications of the training algorithm exist for folding networks as for standard feed-forward networks. We mention just a few methods, the ideas of which can be transferred immediately: integration of prior knowledge via a penalty term in the quadratic error which penalizes solutions contradicting the prior information or regularization of the network via weight restriction [21, 96]; modification of the architecture during the training or simplification of the final network via pruning less important weights and neurons [77, 92].

One problem is to be dealt with, while training recurrent and folding architectures, which does not occur while training feed-forward networks: the problem of long-term dependencies [16]. Roughly, the problem tells us that the training scenario is numerically ill-behaved if a function is to be learned where entries at the very beginning of long sequences or labels near the leaves of deep trees determine the output of the function. Hence several approaches try to modify the architecture, the training algorithm, or both

[14, 40, 54, 67]. This problem has not yet been satisfactorily solved; we will consider some correlated theoretical problems in Chapter 5.

2.3 Background

The above argumentation tells us that a large variety of training methods are available in order to use folding networks in practice. Before describing several areas of application in more detail, we want to mention the origin of the approach. Despite the success of neural networks in several different areas dealing with continuous data [25, 121], their ability to process symbolic data was doubted in [34], for example. One of the main criticisms is that a natural representation of symbolic data in a finite dimensional vector space does not exist because symbolic data is highly structured and consists of an *a priori* unlimited length. Since standard networks only deal with distributed representations of the respective data in a finite dimensional vector space, they cannot be used for processing symbolic data. In reaction to the criticism [34], several approaches including the RAAM and LRAAM [101, 122, 124], tensor constructions [117], BoltzCONS [128], and holographic reduced representations [100] were proposed as mechanisms to deal with some kind of structured data. One key problem is to find a mechanism that maps the structured data into a subsymbolic representation, *i.e.*, a real vector of fixed dimension. For this purpose, connectionistic models are equipped with an appropriate recurrent dynamics in the above approaches. Actually, the RAAM and LRAAM already contain the same dynamics as folding networks, but training proceeds in a different way. They consist of two parts, the first of which is used for recursive encoding of trees into a finite dimensional vector space, the second of which performs a recursive decoding of finite dimensional vectors to trees. Formally, the LRAAM consists of two networks f and g with input dimension $m + k \cdot n$ or n, respectively, and output dimension n or $m + k \cdot n$, respectively. It computes $\bar{h}_Y \circ \tilde{g}_y$ where $\bar{h}_Y : \mathbb{R}^n \to (\mathbb{R}^m)_k^*$ is defined by

$$\bar{h}_Y(x) = \begin{cases} \bot & x \in Y \\ h_0(x)(\bar{h}_Y(h_1(x)), \ldots, \bar{h}_Y(h_k(x))) & \text{otherwise}, \end{cases}$$

where $Y \subset \mathbb{R}^n$ and $h = (h_0, h_1, \ldots, h_k)$. Hence h_0 outputs the label, h_1, ..., h_k compute codes for the k subtrees. $\bar{h}$ performs a dual computation compared to $\tilde{g}$. The LRAAM performs some kind of universal encoding of tree structures. The two networks g and h are trained simultaneously such that the composition yields the identity on k-trees. For this purpose, some kind of truncated gradient descent method is used [101]. Afterwards, one can address any learning task that deals with tree structures as inputs or outputs with a standard network, since the composition of a standard network with one or two parts of the LRAAM reduces the learning task to the learning of a mapping between finite dimensional vector spaces. See Fig.2.5 as an illustration.

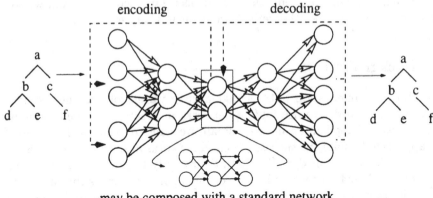

Fig. 2.5. The LRAAM performs both encoding and decoding. Hence composition of these parts with a standard network can be used to learn arbitrary mappings with trees as input or output.

Folding networks deal with only one part of the problem: mappings from tree structures into a real vector space. Hence they use only the encoding part of the LRAAM and combine it directly with a standard network. The training is fitted to the specific learning task. This method allows them to store only the valuable information that is necessary for the specific task and makes the efficient processing of the desired information possible. Additionally, it will turn out in Chapter 4 that folding networks are more likely to succeed from a theoretical point of view compared to the LRAAM.

The other approaches mentioned above have in common the fact that complex data is encoded into a representation as a real vector which is universal and not fitted to the specific learning problem, either. The tensor construction, BoltzCONS, and the holographic reduced representation deal with a fixed encoding which is not learned or modified at all. As already mentioned, RAAM and LRAAM learn the encoding, but only the structure of the input data is relevant for this process – the classification task is not taken into account. In contrast, recurrent and folding networks fit the encoding to the specific learning task.

Actually, this method is already used very successfully in the special case of linear trees, *i.e.*, lists. Recurrent networks are a natural tool in any domain where time plays a role, such as speech recognition and production, control, or time series prediction, to name just a few [39, 43, 80, 87, 91, 126, 140]. They are also used for the classification of symbolic data as DNA sequences [102]. Elman networks are a special case [29, 30], with the main difference that the learning algorithm is only a truncated gradient descent and hence training is commonly less efficient compared to standard methods used for training recurrent networks [54]. Both architectures compete with feed-forward networks which cut the maximum length of an input sequence to a fixed value and can therefore only deal with an *a priori* limited time context [15, 57, 87].

Folding networks have been designed to enable neural networks to classify classical terms, logic formulas, and tree structures so that they can be used in symbolic domains, for the control of search heuristics in automated deduction [41, 111], for example. Furthermore, successful applications exist in various areas.

2.4 Applications

In the following, we have a closer look at some of these applications focusing on the method of how the respective data is encoded by a tree structure.

2.4.1 Term Classification

The set of terms over a finite signature and finite set of variables can be represented by trees in the following way: Denote by k the maximum arity of the function symbols, by c some injective function which maps the symbols, variables, constants, and function symbols, to real numbers. A variable X is encoded by the k-tree $C(X) = c(X)(\bot, \ldots, \bot)$, a term $f(t_1, \ldots, t_{k_1})$ $(k_1 \le k)$ is encoded by the k-tree $C(f(t_1, \ldots, f_{k_1})) = c(f)(C(t_1), \ldots, C(t_{k_1}), \bot, \ldots, \bot)$. Reasonable training tasks are to learn mappings $f : \{t \in \mathbb{R}_k^* \mid t = C(T)$ for some term $T\} \to \{0,1\}$ where $f(C(T)) = 1 \Leftrightarrow T'$ is a subterm of T for some fixed term T', or $f(C(T)) = 1 \Leftrightarrow T$ is an instance of a fixed term T' containing variables, or $f(C(T)) = 1 \Leftrightarrow$ a Boolean combination of similar characteristics as above holds, ...

Results for these kinds of training problems can be found in [42, 73]. Most of the considered tasks can be solved with folding networks $h \circ \tilde{g}_y$ where h and g do not possess a hidden layer, with standard training methods, and more than 93% correct classified data on a test set.

2.4.2 Learning Tree Automata

A (finite, bottom-up, deterministic) tree automaton is a tuple $(\Sigma, Q, b, F, \delta)$ where Σ, the input alphabet, and Q, $(Q \cap \Sigma = \emptyset)$, the set of states, are finite sets, $b \in Q$ is the initial state, $F \subset Q$ is the nonempty set of final states, and $\delta : \Sigma \times Q^k \to Q$ is the transition function. A tree $t \in \Sigma_k^*$ is accepted by the automaton if and only if $\tilde{\delta}_b(t) \in F$. Assuming $\Sigma \subset \mathbb{R}$, natural tasks for a folding architecture are to learn the function $f : (\Sigma_k)^* \to \{0,1\}$ with $f(t) = 1$ if and only if the tree t is accepted by some specified tree automaton.

In [72], folding networks are successfully trained on tasks of this kind with standard training methods; in all cases an accuracy of nearly 100% is obtained on a test set. Furthermore, their in-principle ability of simulating

tree automata is established. Moreover, training turns out to be successful (*i.e.*, more than 94% accuracy on a test set) for languages that are not
recognizable by tree automata either. [72] reports very good results for the
recognition of the language $\{t \in \{f, a, b\}_2^* \mid t$ contains an instance of the term
$f(X, X)$ as subterm$\}$. We will deal with correlated problems concerning the
computational capability of folding networks in Chapter 3.

2.4.3 Control of Search Heuristics for Automated Deduction

Several calculi in automated theorem proving, for example, the resolution
calculus or model elimination [81, 104], try to resolve a goal φ successively
with a set of formulas M in order to see whether φ follows from M or not. The
proof process can be modeled by a search tree with nodes labeled with the
actual states of the theorem prover and sons of the respective nodes which
represent the possible proof steps. If φ follows from M, then at least one
path in this tree leads to a valid proof. See Fig. 2.6 as an example for such
a proof tree. Obviously, the success of the theorem prover depends on the
order according to which the single possible states are explored. Since the
fan-out of the nodes in the search tree is typically larger than one at every
proof step, a naive search leads to an exponential amount of time. In order
to prevent this fact, one can try to find heuristics telling us which state is to
be visited first. Formally, we want to find a mapping from the single nodes
of the search tree into the real numbers such that a search visiting the states
in accordance with the induced ranking leads to a proof in a short time.

The single states of the theorem prover consist of a finite set of logical
formulas and hence can be represented by a finite set of trees in a natural
way. Therefore folding networks can be used in order to learn the mapping
of finite sets of trees into the real numbers which represents an appropriate
ranking. In [41], this method is applied to several word problems from group
theory. The training data is obtained from several different (truncated) proof
trees where the states on a proof path are mapped to high values and states
on a failure path are mapped to low values in a first approximation. Since the
same state may occur on a proof path as well as on a failure path, this naive
approach does not work. Hence in [41] the quadratic error is modified such
that the error is small if at least one state on every failure path is ranked
with a lower value than all states on at least one proof path. The use of
this modified error function training yields very good results: 17 of 19 proofs
which could not be obtained within reasonable time without a ranking could
be performed with the ranking which was learned with a folding network and
standard training methods.

2.4.4 Classification of Chemical Data

In [110], folding networks are used compared to standard feed-forward networks for the task of mapping chemical data (here: triazines) to their activity

Set of formulas:
l(a,b). l(b,c). l(e,c).
L(X,X).   L(X,Y):-L(Y,X).   L(X,Y):-l(X,Z),L(Z,Y).

The goal L(a,e) leads to the following search tree:

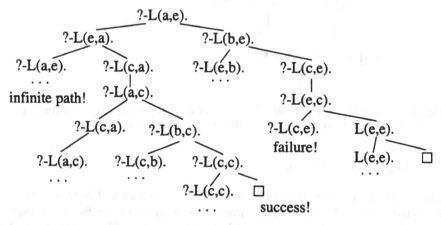

Fig. 2.6. Example for a search tree of an automated theorem prover: The question whether $L(a, e)$ follows from the given set leads to a search tree with several proof paths, but several failure and infinite paths as well.

(here: their ability to inhibit the reproduction of cancer cells). For this purpose the chemical data is represented by terms in an appropriate way. A typical structure is depicted in Fig. 2.7. Two basic phenyl rings can be found in every structure of the training set, the second of which possesses at positions 3 and 4 variable structures: atoms, basic molecules, another ring, or a bridge, as an example. Hence a representation by a binary symbol ring(_,_) which captures the basic ring structure is possible. The two places indicate that at positions 3 and 4 variable structures may be found. These are represented by real values if we deal with atoms or basic molecules, by the symbol ring(_,_), if another ring is found, or bridge(_,_), if a bridge takes place. In the latter case, the first place represents the structure of the bridge itself and the second place represents the free position. Hence we obtain a term representation, for example, the term ring(CL,bridge(CH_2CH_2,_)) for the structure in Fig. 2.7. In addition to the single symbols, physio-chemical attributes like polarizability may be encoded in the labels of the corresponding tree. Training on these data with standard methods yields a folding network such that the spear-man's correlation (*i.e.*, the correlation of the ranking induced by the network) is improved from 0.48 for feed-forward networks to 0.57 for folding networks on a test set. Note that both the labels and the structure encode valuable information in this example if physio-chemical attributes are added in the trees representation.

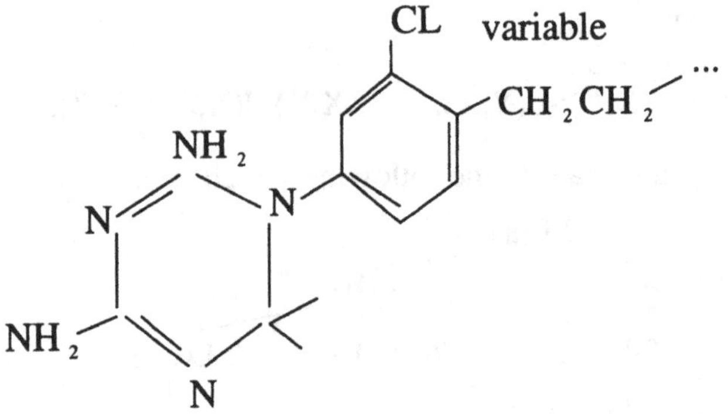

Fig. 2.7. Structure of a triazine: Two phenyl rings form the basic structure, the second of which may possess different structures at positions 3 and 4.

2.4.5 Logo Classification

Images of logos that are subject to noise are to be recognized in an approach presented in [24]. The logos are represented by trees in the following way: First, classical edge detection algorithms extract edges. These are encoded via features like their curvature, length, center point, ... and constitute the labels of the nodes. The nodes are arranged in a tree such that a node becomes the son of another node if the edge represented by the latter node surrounds the former edge by at least 270 degrees. Since this method yields trees with a large fan-out, the approach in [24] applies a method which substitutes the trees by binary trees first. For this data, the approach in [24] reports a test set accuracy of at least 88% for several different settings which are trained with folding networks and standard training algorithms.

Other possible areas of application are reported in [36, 38]. Hence folding networks have turned out to be successful in practice in several different areas. In the following, we investigate a theoretical justification of this fact.

Chapter 3

Approximation Ability

In this chapter we deal with the ability of folding networks to approximate a function from structured data into a real vector space in principle. In a concrete learning task some empirical data $(x_i, f(x_i))$ of an unknown function f which is to be learned is presented. We want to find a folding network that maps the data correctly. That is, the inputs x_i are mapped by the network to values approximating $f(x_i)$. Furthermore, this approximating folding network should represent the entire function f underlying the data correctly. That is, it should map any tree x to a value approximating $f(x)$ even if x is different from all training examples.

But we have no chance of finding such a network unless any finite set of data can be correctly approximated by a folding network and, more significantly, unless any function we want to approximate can be approximated in some way by a folding network with an appropriate number of neurons.

Of course, the notation of approximation is to be made more precise. A function can be approximated in the maximum norm, *i.e.*, the network's output differs for any input at most a small value ϵ from the desired output. In particular, trees of arbitrary height are to be predicted correctly. This means that we want to classify any tree correctly even if the tree is so high such that it will rarely occur. However, situations exist where approximation in the maximum norm seems appropriate, *e.g.*, if the long-term behavior of a dynamic system is to be approximated.

Alternatively, we could restrict the approximation to trees that seem reasonable, excluding, for example, trees of a large height. Formally, we demand only that the probability of trees where the network's output differs significantly from the desired output can be made arbitrarily small. This demand corresponds to an approximation in probability. Since trees with more than a certain height become nearly impossible in any probability measure, this notation of approximation restricts the consideration to inputs with only a restricted recurrence.

It would be nice to not only ensure the approximation capability for some – maybe very large – network but to find explicit bounds on the number of hidden layers and neurons that are necessary in such an architecture. Bounds on the number of hidden layers and neurons limit *a priori* the architectures we have to consider unless a correct representation of the empiric al data is found.

The following questions are worth considering as a consequence:

1. Can any finite set of data be approximated or interpolated, re: pectively, by a folding network?
2. Can any reasonable function be approximated in probability?
3. Can any reasonable function be approximated in the maximur i norm?
4. Can the number of neurons and hidden layers that ensure the approximation capability be limited?

The first two questions are answered with 'yes' in section 3: Any fi iite set of data can be interpolated and any measurable function can be appi oximated in probability. Furthermore, to interpolate a set of p patterns, it is sufficient to use a number of neurons which is quadratic in p for an activatioi function like the sigmoidal function. It is sufficient to consider folding netw orks with only a constant number of layers to approximate a measurable fun :tion. On the contrary, the third question is answered with 'no' in general in section 4. But first we introduce some notation.

3.1 Foundations

We start with a formal definition of the term 'approximation' that w consider here.

Definition 3.1.1. *Assume X is a set and $Y \subset \mathbb{R}^n$ for some n. A function $f : X \to Y$ is approximated by a function $g : X \to Y$ in the maximum norm with accuracy ϵ if*

$$\sup_{x \in X} |f(x) - g(x)| \leq \epsilon .$$

Assume X is equipped with a σ-algebra and P is a probability measure on X. A measurable function $f : X \to Y$ is approximated by a measurable function $g : X \to Y$ in probability with accuracy ϵ and confidence δ if

$$P(x \in X \mid |f(x) - g(x)| > \epsilon) \leq \delta .$$

The situation where a finite set of data, i.e., $X \subset \mathbb{R}^n$ is finite, is approximated or even interpolated, i.e., $\epsilon = 0$, is of particular interest whe i dealing with a learning algorithm which tries to fit a network such that it i epresents some empirical data. In the following, $X \subset \mathbb{R}^m$ will be an arbitra :y subset or $X \subset (\mathbb{R}^m)_k^*$ will contain sequences or trees if f is a recursive mapping. Y will be the real vector space $\mathbb{R}^n$ or a compact subset of this space. Any result that leads to a positive or negative fact concerning these two forn alisms is called an approximation result in the following.

For feed-forward networks it is shown in [59] that any continu us function $\mathbb{R}^m \to \mathbb{R}^n$ with compact domain can be approximated arbitr irily well in the maximum norm, and any Borel-measurable function can be approximated arbitrarily well in probability with a network with only one hidden layer with squashing activation function and linear output. Additionally, it is shown in [58] that it is possible to approximate a continuous function in the maximum norm or a measurable function in probability, respectively, with single hidden layer networks and bounded weights if the activation function of the hidden nodes is locally Riemann integrable and nonpolynomial. Furthermore, to interpolate p points with outputs in $\mathbb{R}$ exactly by such a network with unbounded weights it is sufficient to use only p neurons in the hidden layer [119]. For an input dimension m, output dimension n, and continuously differentiable activation with some other conditions, this bound is improved to $2pm/(n+m)$ hidden neurons, which are sufficient for the interpolation of p points in general position [28]. We will use these results in our constructions in order to show some universal approximation property for folding networks. The respective feed-forward networks will be accompanied by a recursive network. The recursive network encodes the respective data in a finite dimensional vector space and hence reduces the approximation problem to a problem for feed-forward networks.

However, as a consequence of these results folding networks with appropriate activation functions can approximate any function of the special form $g \circ \tilde{f}_y : K_k^{\leq t} \to \mathbb{R}^n$, where $t \in \mathbb{N}$ is some maximum input height, $K \subset \mathbb{R}^m$ is a compact set, and g and f are continuous functions. This is due to the fact that g and f can be approximated arbitrarily well on the interesting domains by a feed-forward network. This argumentation is carried out for the case of recurrent networks in [120]. The argumentation leads to bounds on the number of hidden layers: One hidden layer in the recursive part is sufficient and the feed-forward part can even manage without hidden neurons because any nonlinearity can be considered to be part of the recursive function $\tilde{f}_y$. But it does not allow a limitation of the number of neurons necessary to interpolate a finite number of points. Furthermore, it cannot be applied to an arbitrary function which is not written in a recursive form *a priori*.

Note that trees up to a restricted height are considered, therefore the previous argumentation only leads to approximation in probability. Approximation in the maximum norm is possible if the recursive function $g \circ \tilde{f}_y$ which has to be approximated has a special form: If $g \circ \tilde{f}_y : \Sigma_k^* \to \{0,1\}$ starts from a finite alphabet Σ and can be computed in the case of linear inputs with a finite automaton, in the case of trees with a tree automaton, it can then be approximated with a folding network with appropriate, *e.g.*, sigmoidal or perceptron activation function such that the behavior is equivalent to the automaton's behavior on all inputs [22, 72, 97, 127]. It is possible to derive explicit bounds on the number of neurons required for the simulation. It is sufficient to use the same number of hidden neurons as states in the automaton exist. Furthermore, even Mealy-automata or Mealy-tree-automata, respectively, can be simulated by such a network which allows the approxi-

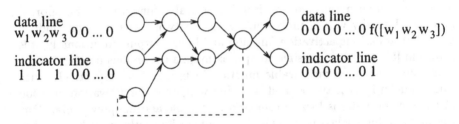

data line
$w_1 w_2 w_3 0 0 \dots 0$

indicator line
$1 \; 1 \; 1 \; 0 \, 0 \dots 0$

data line
$0 \, 0 \, 0 \, 0 \dots 0 \, f([w_1 w_2 w_3])$

indicator line
$0 \, 0 \, 0 \, 0 \dots 0 \, 1$

Fig. 3.1. Recurrent network as a computational model; both, input and output are encoded by two lines indicating the fact that data is present and the data, respectively

mation of some functions with outputs in a finite but not necessarily binary alphabet.

Again, this approach only leads to the approximation of functions in a very special form. In particular, the functions *a priori* have a recursive structure. This property is not known *a priori* in general term classification tasks. Apart from this fact, the results allow the approximation of recursive inputs with unlimited height.

Furthermore, the argumentation gives rise to the question as to whether the simulation of finite automata can be generalized to a simulation of general Turing machines with recurrent networks, too. In order to answer this question it is necessary to define what a computation with a recurrent network looks like. In [116], for example, two possible formalisms are presented. The first formalism operates on binary words w of length at least 1, *i.e.*, on $w \in \{0,1\}^+$. We refer to the single letters in w with the symbols w_i.

Definition 3.1.2. *A recurrent neural network* computes *a function* $f :$ $\{0,1\}^+ \to \{0,1\}$ *(which may be a partial function) on on-line inputs if the network computes* $h \circ \tilde{g}_y : (\mathbb{R}^2)^* \to \mathbb{R}^2$ *such that the following holds*

- *For every word* $w \in \{0,1\}^+$ *where* $f(w)$ *is defined, a number* $t \in \mathbb{N}$, *the computation time, exists with*

$$h \circ \tilde{g}_y([(w_1,1),(w_2,1),\dots,(w_n,1),\underbrace{(0,0),\dots,(0,0)}_{t-|w| \text{ times}}]) = (f(w),1)$$

and for every prefix of this input string with length in $[|w|,t[$ *the network outputs* $(0,0)$.
- *If* $f(w)$ *is not defined,*

$$h \circ \tilde{g}_y([(w_1,1),(w_2,1),\dots,(w_n,1),\underbrace{(0,0),\dots,(0,0)}_{t \text{ times}}]) = (0,0)$$

for all $t \in \mathbb{N}$.

Consequently, two types of input and output neurons exist: one where the input or output values, respectively, of function f can be found, and one which only indicates with a value 1 that data is present and with a value 0 that the network is still computing. See Fig. 3.1.

Another formalization of a computation with a recurrent network gets the input information from the activation value of one specified neuron which encodes the input w. After reading this input, the network is allowed to compute for some time unless the output can be found analogous to the first formalism. Since no further input information is necessary, but the computation may take some time, such a network formally works on sequences with elements in a 0 dimensional vector space $\mathbb{R}^0$. We denote the dummy element in $\mathbb{R}^0$ by $\top$.

Definition 3.1.3. *Assume* $c : \{0,1\}^+ \to \mathbb{R}$ *is an injective function which serves as an encoding function.* $f : \{0,1\}^+ \to \{0,1\}$ *is computed by a recurrent network on off-line inputs which are encoded via* c *if a recurrent architecture* $h \circ \tilde{g}_{(\_,\mathbf{y})}$ *exists, where only one coefficient of the initial context is not specified, such that for every word* $w \in \{0,1\}^+$ *and the function* $h \circ \tilde{g}_{(c(w),\mathbf{y})} : (\mathbb{R}^0)^* \to \mathbb{R}^2$ *the following holds:*

- *For any word* $w \in \{0,1\}^+$, *where* $f(w)$ *is defined, a number* $t \in \mathbb{N}$, *the computation time, exists with* $h \circ \tilde{g}_{(c(w),\mathbf{y})}(\underbrace{[\top,\ldots,\top]}_{t\,\text{times}}) = (f(w),1)$ *and for*

  every prefix of this input string with shorter length the network $h \circ \tilde{g}_{(c(w),\mathbf{y})}$ *outputs* $(0,0)$.
- *If* $f(w)$ *is not defined,* $h \circ \tilde{g}_{(c(w),\mathbf{y})}(\underbrace{[\top,\ldots,\top]}_{t\,\text{times}}) = (0,0)$ *for every* $t \in \mathbb{N}$.

As in the first formalism two output neurons exist: one for the data and one that indicates whether data is present. But the input is encoded in the initial value of one neuron directly using c. All input neurons of the recurrent network are dropped. c can be chosen as $c([w_1,\ldots,w_n]) = \sum_{i=1}^{n}(2w_i+1)/(4^i)$, for example. This encoding is used in [116].

Obviously, the demand for an output of exactly 0 or 1, respectively, can be modified to demand values near 0 or 1. This is appropriate in the case of real valued outputs if we deal with a sigmoidal output activation, for example. Taking t as the computation time, the notion of a computation in linear, polynomial, or exponential time is defined in both of the above formalisms.

In [116] it is shown that any Turing machine can be simulated by a recurrent network with semilinear activation in the on-line mode and in the off-line mode. The simulation only leads to a linear time delay. Therefore it follows that any mapping which can be computed by a Turing machine in constant time can be interpolated in the maximum norm by a recurrent network with semilinear activation. The number of neurons which is sufficient for such an approximation depends on the Turing machine which computes the mapping. A simulation result for the standard sigmoidal function also exists [66]. But no approximation results can be derived from this simulation because it only works for off-line inputs and requires exponential computation time.

Another approach addressing the computational capability of recurrent networks is [115]. There it is shown that nonuniform Boolean circuits can be simulated by recurrent networks with semilinear activation. Nonuniform Boolean circuits here mean an arbitrary family $(B_n)_{n \in \mathbb{N}}$ of Boolean circuits B_n, where B_n has n input nodes and computation nodes with a function from {AND, OR, NOT} from which one node is specified as the output. The computation time of a simulating network is polynomially correlated to the depth of the circuits. The number of neurons necessary for the simulation is fixed. This simulation demonstrates that recurrent networks as a computational model can compute every function in exponential time, because every function can be computed by appropriate nonuniform Boolean circuits. Furthermore, in [115] the other direction, the simulation of recurrent networks with nonuniform Boolean circuits is also considered. It is shown that any recurrent network with an activation function with bounded range that is Lipschitz continuous in a neighborhood of every point, *e.g.*, the standard sigmoidal function, can be computed by nonuniform Boolean circuits with resources polynomially correlated to the computation time of the network. In particular, any function that cannot be computed by nonuniform circuits with polynomial resources cannot be approximated by a recurrent network in the maximum norm – a negative result concerning our third question at the beginning of this chapter. Another demonstration of the computational capability of recurrent networks is given in [114]. Another proof of the Turing universality can be found in [68].

Although all these simulations demonstrate the computational power of recurrent networks, they depend on the fact that the activations of the neurons are real numbers where an arbitrary amount of data can be stored and the computation can be performed with arbitrary precision. Otherwise, *e.g.*, for simple perceptron networks, the internal stack would be limited. In order to take into account that in real computations the internal stack representation may be partially disturbed, recurrent networks are investigated in [84], where the computation is subject to some noise. If the noise is limited the approximation of finite automata is still possible with noisy networks but Turing machines can no longer be simulated. Any network affected by a piecewise equicontinuous noise process can only recognize regular languages. The situation becomes even worse if the noise is not limited, *e.g.*, Gaussian. [85] deals with this scenario. It is shown that any language L that can be accepted by such a network has the property that the fact $w \in L$ only depends on the last k values in the word w for a fixed number k. This result reduces the power to deal with recursive data considerably. In fact, it shows that simple feedforward networks with an *a priori* restricted recurrence of the inputs have the same computational capabilities as recurrent networks if the computation is noisy.

3.2 Approximation in Probability

As already mentioned in the previous section, approximation results for some special functions which are written *a priori* in a recursive form exist. In contrast, we will consider the situation as it usually occurs in recursive learning tasks where symbolic data is to be classified: An arbitrary function $f : (\mathbb{R}^m)_k^* \to \mathbb{R}^n$ is to be learned if some examples of f are present. Consequently, an arbitrary function is to be approximated by a folding network.

3.2.1 Interpolation of a Finite Set of Data

First we show that every finite set of data can be approximated with accuracy $\epsilon = 0$, hence it can be interpolated. In symbolic learning tasks the complex terms are made up of symbols from a finite alphabet. Therefore, a finite number of terms from Σ_k^* with a finite alphabet Σ is to be interpolated in this case. We can assume that the elements of Σ are encoded by natural numbers.

Lemma 3.2.1. *Assume* $\Sigma = \{a_1, \ldots, a_s\} \subset \mathbb{N} \backslash \{0, 1\}$ *is a finite alphabet,* $k \in \mathbb{N}$, t_1, ..., $t_p \in \Sigma_k^*$ *is a finite set of p terms, and* $f : \Sigma_k^* \to \mathbb{R}^n$ *is a function. There exists a folding network* $h \circ \tilde{g}_0 : \Sigma_k^* \to \mathbb{R}^n$ *which interpolates the finite set of data, i.e.,* $f(t_i) = h \circ \tilde{g}_0(t_i)$ *holds for all* t_i.
    The network can be chosen as follows: The part $g : \Sigma \times \mathbb{R}^{2k} \to \mathbb{R}^2$ *consists of* $2(k - 1)$ *multiplying hidden neurons and 2 outputs with linear activation function, the part* $h : \mathbb{R}^2 \to \mathbb{R}^n$ *is an MLP with one hidden layer with* $n \cdot p$ *neurons with squashing or locally Riemann integrable and nonpolynomial activation and linear output neurons.*

Proof. The rough idea is as follows: The trees t_i are first encoded into $\mathbb{R}^2$ via $\tilde{g}_0$ and then mapped to $f(t_i)$ via h. If $\tilde{g}_0$ is injective on t_1, ..., t_p, the possibility of interpolating with h follows from interpolation results concerning MLPs.

    Assume each of the numbers a_i needs exactly d digits for a representation with maybe leading digits 0. A unique representation of a tree t is the string representation

$$s_t = \begin{cases} \underbrace{0 \ldots 0}_{d \text{ digits}} & \text{if } t \text{ is the empty tree } \perp, \\ \underbrace{0 \ldots 0}_{d \text{ digits}} 1 \, a_i s_{t^1} \ldots s_{t^k} & \text{if } t \text{ is } a_i(t^1, \ldots, t^k). \end{cases}$$

The scaled number $0.s_t$ can be computed by a recursive function as follows: For two strings $(0.s_1, (0.1)^{\text{length}(s_1)})$, $(0.s_2, (0.1)^{\text{length}(s_2)})$ the concatenation is

$$(0.s_1 s_2, (0.1)^{\text{length}(s_1 s_2)}) =$$
$$(0.s_1 + (0.1)^{\text{length}(s_1)} \cdot 0.s_2, (0.1)^{\text{length}(s_1)} \cdot (0.1)^{\text{length}(s_2)}),$$

consequently $(0.s_t, (0.1)^{\text{length}(s_t)-d})$ can be computed as $(0,0)$ if t is the empty tree, and it can be computed by

$$
\begin{aligned}
(0.0\ldots 01 &+ (0.1)^{2d} \cdot a_i + (0.1)^{2d} \cdot 0.s_{t^1} \\
&+ (0.1)^{2d+\text{length}(s_{t^1})} \cdot 0.s_{t^2} \\
&+ (0.1)^{2d+\text{length}(s_{t^1})+\text{length}(s_{t^2})} \cdot 0.s_{t^3} \\
&+ \cdots \\
&+ (0.1)^{2d+\text{length}(s_{t^1})+\ldots+\text{length}(s_{t^{k-1}})} \cdot 0.s_{t^k}, \\
(0.1)^{2d+\text{length}(s_{t^1})+\ldots+\text{length}(s_{t^k})-d})
\end{aligned}
$$

if $t = a_i(t^1, \ldots, t^k)$. This mapping equals $\tilde{g}_{(0,0)}$, where $g : \Sigma \times \mathbb{R}^{2k} \to \mathbb{R}^2$ is defined by $g(x, x_1, y_1, \ldots, x_k, y_k) = (z_1, z_2)$ with

$$
\begin{aligned}
z_1 &= (0.1)^d \cdot (1 + (0.1)^d x + (0.1)^d x_1 + (0.1)^{2d} y_1 x_2 + \\
&\qquad \ldots + (0.1)^{kd} y_1 \ldots y_{k-1} x_k), \\
z_2 &= (0.1)^{d(k+1)} y_1 \ldots y_k,
\end{aligned}
$$

which can be computed by a network as stated in the lemma.

$\tilde{g}_0$ embeds the trees t_1, $\ldots$, t_p injectively into $\mathbb{R}^2$. The function which maps the finite set of images in $\mathbb{R}^2$ to the values $f(t_i)$ can be completed to a continuous function and therefore be approximated arbitrarily well by an MLP with one hidden layer with activation functions as stated above [58, 59]. From [119] it follows that even np hidden neurons are sufficient for an exact interpolation of the p images in $\mathbb{R}^n$.   □

As a consequence, a network can be found which maps the empirical data correctly in any concrete learning task. Explicit bounds on the number of neurons limit the search space for an interpolating architecture. If a concrete algorithm does not manage to find appropriate weights such that a network with an architecture as described in Lemma 3.2.1 produces a small error on the data, this is not due to the limited capacity of the architecture but to the weakness of the learning algorithm. In particular, situations to test the in principle capability of an algorithm of minimizing the empirical error can be derived from these bounds. However, the architecture in the recursive part is not a common one. But one can substitute it by a standard MLP with a number of neurons correlated to k.

This is due to the fact that every activation function we used in the previous construction can be approximated in an appropriate way by a standard activation function. In the following we will use the fact that the possibility of approximating an activation function enables us to approximate the entire folding network, too, in a formal notation:

Lemma 3.2.2. *Assume $f : \mathbb{R}^{m+k \cdot n} \to \mathbb{R}^n$ is a function which is the composition of continuous functions $f_1 : \mathbb{R}^{n_1 = m+k \cdot n} \to \mathbb{R}^{n_2}$, $f_2 : \mathbb{R}^{n_2} \to \mathbb{R}^{n_3}$, $\ldots$, $f_l : \mathbb{R}^{n_l} \to \mathbb{R}^n$, and every function f_i can be uniformly approximated by f_i^α ($\alpha \to 0$) on compact sets $C_i \subset \mathbb{R}^{n_i}$ with $f_i(C_i) \subset C_{i+1}$, $f_l(C_l) \subset C_1^2$ where*

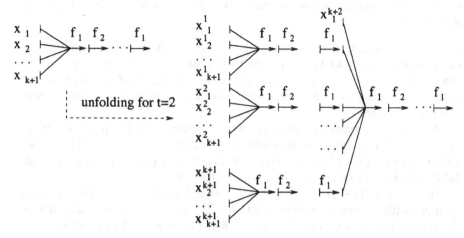

Fig. 3.2. Unfolding of a recursive mapping for an input tree of height 2: The network is copied for each node in the tree

$C_1 = C_1^1 \times (C_1^2)^k$, such that the respective image of C_i under f_i has a positive distance to the boundary of C_{i+1} or C_1^2, respectively. Assume that $T \in \mathbb{N}$. Then the function $\tilde{f}_{\mathbf{y}}$ can be approximated arbitrarily well in the maximum norm for every $\mathbf{y} \in (C_1^2)^k$ on trees in $(C_1^1)_k^{\leq T}$ by a recursive function which is induced by a composition of f_i^α for appropriate α.

Proof. $\tilde{f}_{\mathbf{y}}$ is formally unfolded for trees of height T with the maximum number of labels (see Fig. 3.2). Starting on the right hand side at the last function in this computation we can substitute each function f_i by an approximating function f_i^α with the following property: for inputs x' in C_i, which differ only slightly from a value x, the output $f_i^\alpha(x')$ differs only slightly from $f_i(x)$ such that the difference does not exceed the desired tolerance of the approximation of $\tilde{f}_{\mathbf{y}}$. This argument uses the equicontinuity of f_i on compact sets as well as the uniform approximation, *i.e.*, the property $\forall \epsilon \exists \alpha_0 \forall \alpha_1 \leq \alpha_0 \forall x \in C_i |f_i^{\alpha_1}(x) - f_i(x)| < \epsilon$. Proceeding in the same way for every function that occurs in the unfolded computation, each f_i is substituted by an appropriate f_i^α. Here the tolerance for f_i comes from the already computed tolerance for the inputs of f_{i+1} and of the distance $f_i(C_i)$ from the boundary of C_{i+1}. Note that the f_i occur more than once in the unfolded computation. We can simply choose the minimum α afterwards. $\qquad \square$

This result can be used to substitute the linear and square activation function in the recursive part of the network constructed in Lemma 3.2.1.

Lemma 3.2.3. *Assume the situation is as in Lemma 3.2.1. Then a folding network $h \circ \tilde{g}_{\mathbf{y}}$ exists with $h \circ \tilde{g}_{\mathbf{y}}(t_i) = f(t_i)$ for all i, where h is as in Lemma 3.2.1, $g : \Sigma \times \mathbb{R}^{2k} \to \mathbb{R}^2$ is an MLP with $O(\lg k)$ hidden layers each with*

$O(k)$ neurons with an activation function σ that is locally C^2 with $\sigma'' \neq 0$. y depends on σ.

Proof. The mapping g in Lemma 3.2.1 is to be substituted by a mapping that can be computed by an MLP with activation function σ. Note that g contains the terms y_1, $y_1 \cdot y_2$, ..., $y_1 \cdot \ldots \cdot y_k$, which can be computed in $O(\lg k)$ layers with $O(k)$ neurons in each layer, each unit computing the product of at most two predecessors.

Adding $k + 1$ units in each layer which simply copy the input values x, x_1, ..., x_k of g, adding one layer with $k + 1$ multiplying units, and one layer with two units with identical activation function, g can be computed in an MLP with $O(\lg k)$ layers each containing $O(k)$ units.

Since $x \cdot y = ((x+y)^2 - x^2 - y^2)/2$ we can substitute the multiplying units by units with square activation functions. The order of the bounds remains the same. Since σ is locally C^2 with $\sigma'' \neq 0$, points x_0 and x_1 exist with

$$\lim_{\epsilon \to 0} \frac{\sigma(x_0 + \epsilon x) - \sigma(x_0)}{\epsilon \sigma'(x_0)} = x \quad \text{and}$$

$$\lim_{\epsilon \to 0} \frac{\sigma(x_1 + \epsilon x) + \sigma(x_1 - \epsilon x) - 2\sigma(x_1)}{\epsilon^2 \sigma''(x_1)} = x^2$$

for all x. The convergence is uniform on compact intervals; in particular, we can choose ϵ such that the recursive mapping which results if the square and identical activation functions in g are substituted by the above terms is still injective on the t_i's. Note that the substituting terms consist of an affine combination of σ, consequently an MLP with activation function σ results: the coefficients in the affine combination are considered to be part of the weights of the preceding or following layer, respectively, of the corresponding unit. In the last layer, of course, no following layer exists in g; but due to the recurrence we can put the factors into the weights of the first hidden layer of g and h. It is necessary to change the initial context from $(0,0)$ to $(\sigma(x_0), \sigma(x_0))$. Formally, this method uses the identity $\widetilde{(A \circ f_1)}_y = A \circ (\tilde{f}_2)_{y'}$ which holds every mappings A and f_1 of appropriate arity, vectors $y = A(y')$, and $f_2(x, y_1, \ldots, y_k) = f_1(x, A(y_1), \ldots, A(y_k))$.

We have constructed a mapping that is injective on the t_i's and can be computed with an MLP with activation function σ. Hence a feed-forward network h which maps the images of the t_i to the desired outputs as stated in the theorem exists. $\square$

Consequently, even with a standard architecture the approximation of a finite set of data is possible. In the case of sequences, *i.e.*, $k = 1$, the resources necessary for an interpolation can be further restricted.

Lemma 3.2.4. *Assume $\Sigma \subset \mathbb{N} \backslash \{0, 1\}$ is finite, $f : \Sigma^* \to \mathbb{R}^n$ is a function, and t_1, ..., $t_p \in \Sigma^*$. Then a folding network $h \circ \tilde{g}_0$ exists with $h \circ \tilde{g}_0(t_i) = f(t_i)$ for all i. $g : \Sigma \times \mathbb{R} \to \mathbb{R}$ is a network without hidden units and one output* *with identical activation function. $h : \mathbb{R} \to \mathbb{R}^n$ is a network with one hidden layer with np hidden neurons with squashing or locally Riemann integrable and nonpolynomial activation function and linear outputs.*

The activation function in g can be substituted by any activation function σ which is locally C^1 with $\sigma' \neq 0$. In this case the initial context is $\sigma(x_0)$, where x_0 is some point with $\sigma'(x_0) \neq 0$.

Proof. We assume that exactly d digits are necessary for a representation of $a_i \in \Sigma$. The injective mapping

$$\tilde{g}_0 : \Sigma^* \to \mathbb{R}, \quad [a_{i_1}, a_{i_2}, \ldots, a_{i_t}] \mapsto 0.a_{i_t} a_{i_{t-1}} \cdots a_{i_1}$$

is induced by $g : \Sigma \times \mathbb{R} \to \mathbb{R}$, $g(x_1, x_2) = (0.1)^d x_1 + (0.1)^d x_2$. g can be computed by one neuron with linear activation function. The existence of an appropriate h in the feed-forward part of the recurrent network approximating f follows directly from the approximation and interpolation results in [58, 59, 119].

The linear activation function in g can be substituted by the term $(\sigma(x_0 + \epsilon x) - \sigma(x_0))/(\epsilon \sigma'(x_0))$, which is approximately x for small ϵ. The additional scaling and shifting is considered to be part of the weights of the neuron in the recursive part and h. The initial context is to be changed from 0 to $\sigma(x_0)$. $\square$

The labels often come from a real vector space instead of a finite alphabet if dealing with hybrid data instead of purely symbolic data. The above results can be adapted in this case. However, one additional layer is necessary in order to substitute the real valued inputs by appropriate symbolic values.

Theorem 3.2.1. *Assume $f : (\mathbb{R}^m)_k^* \to \mathbb{R}^n$ is a function and $t_1, \ldots, t_p \in (\mathbb{R}^m)_k^*$. Then a folding network $h \circ \tilde{g}_y$ exists such that $h \circ \tilde{g}_y(t_i) = f(t_i)$ for all i.*

$h : \mathbb{R} \to \mathbb{R}^n$ if $k = 1$ or $h : \mathbb{R}^2 \to \mathbb{R}^n$ if $k = 2$, respectively, is a network with linear output neurons and one hidden layer with np hidden neurons. The activation function of these neurons is either a squashing activation function or some function which is locally Riemann integrable and nonpolynomial. If $k = 1$ then $g : \mathbb{R}^m \times \mathbb{R} \to \mathbb{R}$ is a network with one hidden layer with $O(p^2)$ neurons with a squashing activation function which is locally C^1 with nonvanishing derivative and one output neuron with a locally C^1 activation function with nonvanishing derivative. If $k \geq 2$ then $g : \mathbb{R}^m \times \mathbb{R}^{2k} \to \mathbb{R}^2$ is a network with $O(\lg k)$ layers. The first hidden layer contains $O(p^2)$ neurons with a squashing and locally C^1 activation function with nonvanishing derivative. The remaining hidden layers contain $O(k)$ neurons with a locally C^2 activation function with nonvanishing second derivative.

Proof. The trees $t_1, \ldots, t_p$ are different either because of their structure or because of their labels. We want to substitute the real vectors in the single labels by symbolic values such that the resulting trees are still mutually different. Consider trees $t_1, \ldots, t_{p'}$ of the same structure. At least one coefficient in some label exists in t_1 which is different from the coefficient in the same label in t_2; the same is valid for t_1 and t_3, t_1 and t_4, $\ldots$, $t_{p'-1}$ and $t_{p'}$. Hence we can substitute all coefficients arbitrarily for all but at most $(p'+1)^2$ different values such that the resulting trees are still mutually different. Denote the real values which distinguish the trees $t_1, \ldots, t_p$ by a_i^j, j denoting the coefficient in which the value occurs. Note that the values add up to at most $P = (p+1)^2$. The number of coefficients occurring in dimension i is denoted by P_i.

Now a layer is added to the function g in Lemma 3.2.3 or 3.2.4, respectively, which substitutes the labels by a symbolic value in $\{2, \ldots, P^m + 1\}$, m being the dimension of the labels. More precisely, g is combined with g_1 where

$$g_1(x_1, \ldots, x_m) = 2 + \sum_{i=1}^{m} P^{i-1} \sum_{j=1}^{P_i} (j-1) \cdot 1_{x_i = a_j}(x_i).$$

g_1 computes a representation of the coefficients restricted to the values a_i to base P. Combined with the function g in Lemma 3.2.1 or 3.2.4, respectively, i.e.,

$$g_2(x_1, \ldots, x_m, y_1, z_1, \ldots, y_k, z_k) = g(g_1(x_1, \ldots, x_m), y_1, z_1, \ldots, y_m, z_m)$$

or

$$g_2(x_1, \ldots, x_m, y) = g(g_1(x_1, \ldots, x_m), y),$$

respectively, this yields a prefix representation of the trees such that the images of the trees t_i are mutually different. g_1 can be computed by one hidden layer with the perceptron activation function and linear neurons copying y_1, $\ldots$ or y, respectively, and one linear output neuron: for this purpose, $1_{x_i = a_i}$ is substituted by $H(x_i - a_j) + H(-x_i + a_j) - 1$. The linear outputs of g_1 can be integrated into the first layer of g.

Since we are dealing with a finite number of points, the biases in the perceptron neurons can be slightly changed such that no activation coincides with 0 on the inputs t_i. Hence the perceptron neurons can be approximated arbitrarily well by a squashing function σ because of $H(x) = \lim_{\epsilon \to \infty} \sigma(x/\epsilon)$ for $x \neq 0$. The remaining activation functions can be approximated by difference quotients using activation functions as described in the theorem. The approximation can be performed such that the function is injective on the inputs t_i. Hence a function h mapping the images of the t_i to the desired outputs as described in the theorem exists. $\square$

3.2.2 Approximation of a Mapping in Probability

These results show that for standard architectures interpolation of a finite number of points is possible. This interpolation capability is a necessary condition for a learning algorithm which minimizes the empirical error to succeed.

However, if we want to approximate the entire function f the question arises as to whether the entire function can be represented by such a network in some way. A universal approximation capability of folding networks is a necessary condition for a positive answer to this question. Of course, the simple possibility of representing the function in some way does not imply that any function with small empirical error is a good approximation of f as well, it is only a necessary condition; we will deal with sufficient conditions for a learning algorithm to produce the correct function in the next chapter.

The capability of approximating a finite number of points implies the capability of approximating any mapping $f : \Sigma_k^* \to \mathbb{R}^n$ in probability immediately if Σ is a finite alphabet.

Theorem 3.2.2. *Assume $\Sigma \subset \mathbb{N} \backslash \{0, 1\}$ is a finite alphabet and P is a probability measure on Σ_k^*. For any $\delta > 0$ and function $f : \Sigma_k^* \to \mathbb{R}^n$ there exists a folding network $h \circ \tilde{g}_{\mathbf{y}} : \Sigma_k^* \to \mathbb{R}^n$ such that $P(x \in \Sigma_k^* | f(x) \neq h \circ \tilde{g}_{\mathbf{y}}(x)) \leq \delta$. $\mathbf{y}$ depends on the activation function of the recursive part.*

h can be chosen as an MLP with input dimension 1 if $k = 1$ or input dimension 2 if $k \geq 2$ with linear outputs and a squashing or locally Riemann integrable and nonpolynomial activation function in the hidden layers.

g can be chosen as a network with only 1 neuron with an activation function which is locally C^1 with $\sigma' \neq 0$ in the case $k = 1$. For $k \geq 2$ the function g can be chosen as a network with $O(k)$ neurons with linear and multiplying units or as an MLP with $O(\lg k)$ layers, each containing $O(k)$ units, and an activation function which is locally C^2 with $\sigma'' \neq 0$.

Proof. Since Σ_k^* can be decomposed into a countable union of subsets S_i, where S_i contains only trees of one fixed structure, we can write $1 = P(\Sigma_k^*) = \sum_i P(S_i)$. Consequently, we can find a finite number of tree structures such that the probability of the complement is smaller than δ. Any folding network which interpolates only the finite number of trees of these special structures approximates f in probability with accuracy 0 and confidence δ. A folding network interpolating a finite number of trees with an architecture as stated in the theorem exists because of the approximation lemmata we have already proven.   □

We have shown that folding networks have the capacity to handle mappings that take symbolic terms as inputs. However, recurrent networks are often used for time series prediction where the single terms in the time series are real data, *e.g.*, market prices or precipitation. Even when dealing with trees it is reasonable that some of the data is real valued, especially when symbolic as well as sub-symbolic data are considered, *e.g.*, arithmetical terms which contain variables and are only partially evaluated to real numbers, or image data which may contain the relative position of the picture objects as well as the mean grey value of the objects. Consequently, it is an interesting question whether mappings with input trees with labels in a real vector space can be approximated as well.

Theorem 3.2.3. *Assume P is a probability measure on $(\mathbb{R}^m)_k^*$, $f : (\mathbb{R}^m)_k^* \to \mathbb{R}^n$ is a measurable function, and δ and ϵ are positive values. Then a folding network $h \circ \tilde{g}_y : (\mathbb{R}^m)_k^* \to \mathbb{R}^n$ exists such that $P(x \in (\mathbb{R}^m)_k^* \| f(x) - h \circ \tilde{g}_y(x)| > \epsilon) \leq \delta$. h and y can be chosen as in Theorem 3.2.2. Compared to Theorem 3.2.2, g contains one additional layer with a squashing activation function which is locally C^1 with $\sigma' \neq 0$. The number of neurons in this layer depends on the function f.*

Proof. It follows from [59] that any measurable function $f|\{$trees of a fixed structure with i nodes$\}$ can be approximated with accuracy $\epsilon/2$ and confidence $(\delta \cdot (0.5)^{i+1})/(2 \cdot A(i))$ in the probability induced by P, where $A(i)$ is the finite number of different tree structures with exactly i nodes. Consequently, the entire mapping f can be approximated with accuracy $\epsilon/2$ and confidence $\delta/2$ by a continuous mapping. Therefore we can assume w.l.o.g. that f itself is continuous.

We will show that f can be approximated by the composition of a function $f_1 : \mathbb{R}^m \to \Sigma$ which scans each real valued label into a finite alphabet $\Sigma \subset \mathbb{N}$ and therefore induces an encoding of the entire tree to a tree in Σ_k^* and a function $f_2 : \Sigma_k^* \to \mathbb{R}^n$ which can be approximated using Theorem 3.2.2. Since we are dealing with an infinite number of trees the situation needs additional argumentation compared to Theorem 3.2.1.

A height T exists such that trees higher than T have a probability smaller than $\delta/4$. A positive value B exists such that trees which contain at least one label outside $]-B, B[^n$ have a probability smaller than $\delta/4$. Each restriction $f|S$ of f to a fixed structure S of trees of height at most T and labels in $[-B, B]^n$ is equicontinuous. Consequently, we can find a constant ϵ_S for any of these finite number of structures such that $|(f|S)(t_1) - (f|S)(t_2)| < \epsilon/2$ for all trees t_1 and t_2 of structure S with $|t_1 - t_2| < \epsilon_S$. Take $\epsilon_0 = \min_S\{\epsilon_S\}$. Decompose $]-B, B[$ into disjoint intervals

$$I_1 =]-B, b_1[, \ I_2 =]b_1, b_2[, \ \ldots, I_q =]b_{q-1}, B[$$

of diameter at most $\epsilon_0/\sqrt{m2^T}$ such that $P(t \mid t$ is a tree with a label with at least one coefficient $b_i) < \delta/4$. Note that for two trees of the same structure, where for any label the coefficients of both trees are contained in the same interval, the distance is at most ϵ_0, *i.e.*, the distance of the images under f is at most $\epsilon/2$. The mapping $f_1 : \mathbb{R}^m \to \Sigma = \{2, \ldots, q^m + 1\}$,

$$(x_1, \ldots, x_m) \mapsto 2 + \sum_{i=1}^{m} q^{i-1} \sum_{j=1}^{q} (j-1) \cdot 1_{I_j}(x_i)$$

encodes the information about which intervals the coefficients of **x** belong to. Here 1_{I_j} is the characteristic function of the interval I_j. f_1 gives rise to a mapping $(\mathbb{R}^m)_k^* \to \Sigma_k^*$, where each label in a tree is mapped to Σ via f_1. Therefore we can define $f_2 : \Sigma_k^* \to \mathbb{R}^n$, which maps a tree $t \in \Sigma_k^*$ to $f(t')$, where $t' \in (\mathbb{R}^m)_k^*$ is a fixed tree of the same structure as t with labels in the corresponding intervals that are encoded in t via f_1.

f_2 can be interpolated on trees up to height T with a mapping $h \circ \tilde{g}_{1y} : \Sigma_k^* \to \mathbb{R}^n$, as described in Lemma 3.2.1. Because of the construction, the mapping $h \circ \tilde{g}_y : (\mathbb{R}^m)_k^* \to \mathbb{R}^n$, where $g(\mathbf{x}, \mathbf{y}^1, \ldots, \mathbf{y}^k) = g_1(f_1(\mathbf{x}), \mathbf{y}^1, \ldots, \mathbf{y}^k)$ for $\mathbf{x} \in \mathbb{R}^m$, $\mathbf{y}^i \in \mathbb{R}^2$ differs at most $\epsilon/2$ from f for a set of probability at least $1 - 3\delta/4$. The characteristic function 1_{I_i} in f_1 can be implemented by two neurons with perceptron activation function because $1_{I_i}(x) = \mathrm{H}(x - b_{i-1}) + \mathrm{H}(b_i - x) - 1$, where $b_0 = -B$ and $b_q = B$. Consequently, we can approximate the identity activation in g via the formula $x = \lim_{\epsilon_1 \to 0}(\sigma(x_0 + \epsilon x_1) - \sigma(x_0))/(\epsilon_1 \sigma'(x_0))$ for a locally C^1 activation function σ and the perceptron activation in g via the formula $\mathrm{H}(x) = \lim_{\epsilon_1 \to 0} \sigma(x/\epsilon_1)$ for a squashing activation function σ. The latter convergence is uniform, except for values near 0. The feed-forward function h approximates a continuous function in the maximum norm such that we obtain a tolerance for the relevant inputs of h such that input changes within this tolerance change the output by at most $\epsilon/2$. Therefore, the approximation process in g results in a mapping which differs at most ϵ on trees of probability at least $1 - \delta$ from the mapping f.

Obviously, an analogous argumentation based on Lemma 3.2.4 leads to the smaller encoding dimension 1 if $k = 1$. $\qquad\qquad\square$

We can conclude that general mappings on trees with real labels can be approximated. Unfortunately, the results lead to bounds on the number of hidden layers, but not on the number of neurons required in the recursive or feed-forward part in the general situation. The number of neurons depends somehow on the smoothness of the function to be approximated. This is due to the fact that the number of representative input values for f, *i.e.*, intervals we have to consider in a discretization, depends on the roughness of f.

In fact, by using this general approximation result the bound for the number of hidden layers can be further improved. Together with the universal approximation we have constructed an explicit recursive form with continuous transformation function for any measurable mapping on recursive data. The approximation results for feed-forward networks can be used to approximate the transformation function, and the result of [120] follows in fact for any measurable function with trees as inputs which is not written *a priori* in a recursive form.

Corollary 3.2.1. *Assume P is a probability measure on $(\mathbb{R}^m)_k^*$. For any measurable function $f : (\mathbb{R}^m)_k^* \to \mathbb{R}^n$ and positive values ϵ and δ there exists a folding network $h \circ \tilde{g}_y : (\mathbb{R}^m)_k^* \to \mathbb{R}^n$ with $P(x \,||\, f(x) - h \circ \tilde{g}_y(x)| > \epsilon) \le \delta$.*

h can be chosen as a linear mapping. g can be chosen as a neural network without hidden layer and a squashing or locally Riemann integrable and nonpolynomial activation function.

Proof. As already shown, f can be approximated by a folding network $f_1 \circ \tilde{f}_{2y}$ with continuous functions f_1 and f_2 computed by feed-forward networks with sigmoidal activation function, for example. For an approximation in probability it is sufficient to approximate $f_1 \circ \tilde{f}_{2y}$ only on trees up to a fixed height with labels in a compact set. Adding n neurons, which compute the output values of f_1, to the outputs of f_2 in the encoding layer if necessary, we can assume that f_1 is linear. f_2 can be approximated arbitrarily well by a feed-forward network with one hidden layer with squashing or locally Riemann integrable and nonpolynomial activation function on any compact set in the maximum norm [58, 59]. In particular, f_2 can be approximated by some g of this form such that the resulting recursive function $\tilde{g}_y$ differs at most ϵ on the trees up to a certain height with labels in a compact set. We can consider the linear outputs of this approximation to be part of the weights in the first layer or the function f_1, respectively, which leads to a change of the initial context y. With the choice $h = f_1$, a network of the desired structure results.   □

This result minimizes the number of layers necessary for an approximation and is valid for any squashing function even if it is not differentiable, *e.g.*, the perceptron function. But the encoding dimension increases if the function that is to be approximated becomes more complicated. A limitation of the encoding dimension is possible if we allow hidden layers in the recursive and feed-forward part.

Corollary 3.2.2. *For any measurable function $f : (\mathbb{R}^m)_k^* \to \mathbb{R}^n$ a folding network $h \circ \tilde{g}_y$ exists which approximates f in probability. The encoding dimension can be chosen as 2. h and g can be chosen as multilayer networks with one hidden layer, locally Riemann integrable and nonpolynomial or squashing activation functions in the hidden layer of g and h, linear outputs in h and a locally homeomorphic output activation function in g (i.e., there exists a nonempty open set U such that $\sigma|U$ is continuous, $\sigma|U : U \to \sigma(U)$ is invertible, and the inversion is continuous, too).*

Proof. As before, f is approximated with $f_1 \circ \tilde{f}_{2y}$. A shift and scaling is introduced in the output layer of f_2 such that the relevant range of $\tilde{f}_{2y}$ – which was a neighborhood of $(0,0)$ by construction – now coincides with the range $\sigma(U)$, where the output activation function σ is homeomorphic. As before, f_1 and $((\sigma|\sigma(U))^{-1}, (\sigma|\sigma(U))^{-1}) \circ f_2$ can be approximated in the maximum norm with feed-forward networks h and $\bar{g}$ such that h and $g = (\sigma, \sigma) \circ \bar{g}$ fulfill the conditions as stated in the corollary.   □

Here again, we can choose the encoding dimension as 1 if $k = 1$. Furthermore, we can limit the weights in the approximating networks in Corollaries 3.2.1 and 3.2.2 by an arbitrary constant $B > 0$ if the activation functions in the networks are locally Riemann integrable, because in this case the weights in the corresponding feed-forward networks can be restricted [58]. The possibility of restricting the output biases in the feed-forward and recursive part depends on the respective activation function. For standard activation functions like sgd a restriction is possible.

3.2.3 Interpolation with $\sigma = H$

Note that although Corollary 3.2.1 also holds for the perceptron activation function, no bounds on the encoding dimension can be derived. It would be nice to obtain an explicit bound for the number of neurons in the perceptron case, too, at least if only a finite number of examples has to be interpolated. Unlike in the real valued case, the number of neurons in the encoding part necessarily increases with the number of patterns since the information that can be stored in a finite set of binary valued neurons is limited. In fact, a brute force method, which simply encodes the trees directly with an appropriate number of neurons, leads to a bound which requires the same number of neurons as nodes in the trees in the encoding part.

Theorem 3.2.4. *Assume Σ is a finite alphabet, $t_1, \ldots, t_p \in \Sigma_k^*$ are input trees, and $f : \Sigma_k^* \to \mathbb{R}^n$ is a function. There exists a folding network $h \circ \tilde{g}_y : \Sigma_k^* \to \mathbb{R}^n$ which interpolates f on these examples t_i.*

h can be chosen as an MLP with one hidden layer with np hidden neurons with squashing activation and linear outputs, g can be chosen as an MLP with squashing activation, one hidden layer and one output layer, each with a number of neurons that is exponential in the maximum height of the input trees and in k if $k \geq 2$ and with a number of neurons that is linear in the length of the sequences for $k = 1$.

Proof. Assume that T is the maximum height of the trees $t_1, \ldots, t_p$. Assume $\Sigma \subset \mathbb{N}$ has s elements. Trees up to height T can be encoded with $d = k^T(4+s)$ binary numbers in the following way: The labels are encoded in s digits in a unary notation $a_i \mapsto 0 \ldots 010 \ldots 0$, the empty tree is encoded as 01110, and any other tree is encoded as

$$a_i(t_1, \ldots, t_k) \mapsto 0110 \, \text{code}(a_i) \, \text{code}(t_1) \ldots \text{code}(t_k).$$

In a network implementation we additionally store the length in a unary way in d digits: $\underbrace{1 \ldots 1}_{\text{length}} 0 \ldots 0$. Starting with the empty tree

$$(011100 \ldots 0, 111110 \ldots 0)$$

the encoding is induced by $g : \mathbb{N} \times \{0,1\}^{2kd} \to \{0,1\}^{2d}$,

$$(a_i, \mathbf{y}^1, \mathbf{l}^1, \ldots, \mathbf{y}^k, \mathbf{l}^k)$$
$$\mapsto (0, 1, 1, 0, 1_{a_i = a_1}, \ldots, 1_{a_i = a_s}, \bar{\mathbf{y}}^1, \ldots, \bar{\mathbf{y}}^k, \text{code of } 4 + s + l^1 + \ldots + l^k),$$

where l^i is the decimal number corresponding to the unary representation $\mathbf{l}^i$, and $\bar{\mathbf{y}}^i$ are the first l^i coefficients of $\mathbf{y}^i$. *i.e.*, the jth neuron in the output of g computes for $j' = j - 4 - s > 0$ the formula $y_j^1 \vee (y_1^2 \wedge l^1 = j' - 1) \vee (y_2^2 \wedge l^1 = j' - 2) \vee \ldots \vee (y_d^2 \wedge l^1 = j' - d) \vee (y_1^3 \wedge l^1 + l^2 = j' - 1) \vee (y_2^3 \wedge l^1 + l^2 = j' - 2) \vee \ldots \vee (y_d^3 \wedge l^1 + \ldots + l^{k-1} = j' - d)$. The test $l^1 + \ldots + l^i = j$ can be performed – with brute force again – just testing every possible tuple $(l^1 = 1 \wedge l^2 = 1 \wedge \ldots \wedge l^i = j - (i - 1)) \vee \ldots$, where $l^i = j$ means to scan the pattern 10 at the places $j, j + 1$ in the unary representation $\mathbf{1}^i$ of l^i. The same brute force method enables us to compute the new length $l^1 + \ldots + l^k$ with perceptron neurons. g can be computed with a network with perceptron neurons where the number is exponential in the maximum height T. Of course, the perceptron functions can be substituted by a squashing activation function σ using the equation $H(x) = \lim_{\epsilon_0 \to 0} \sigma(x/\epsilon_0)$ for $x \neq 0$.

The encoding can be simplified in the case $k = 1$: A sequence of the form $[a_{i_1}, a_{i_2}, \ldots, a_{i_n}]$ is encoded as $\text{code}(a_{i_n}) \ldots \text{code}(a_{i_1})$. This is induced by a mapping which scans the code of the label into the first s units and shifts the already encoded part exactly s places. A network implementation only requires a number of neurons which is linear in T.

The existence of an appropriate h follows because of [58, 59, 119]   □

This direct implementation gives bounds on the number of neurons in the perceptron case, too, although they increase exponentially in T, as expected. However, this enables us to construct situations where the possibility of a training algorithm of minimizing the empirical error can be tested. At least for an architecture as described above a training algorithm should be able to produce small empirical error. If the error is large the training gets stuck in a local optimum.

3.3 Approximation in the Maximum Norm

All approximation results in the previous section only address the interpolation of a finite number of points or the approximation of general mappings in probability. Hence the height of the trees that are to be considered is limited. Consequently the recurrence for which approximation takes place is restricted. In term classification tasks this is appropriate in many situations: In automated theorem proving a term with more than 500 symbols will rarely occur, for example.

3.3.1 Negative Examples

However, when dealing with time series we may be interested in approximating the long-term behavior of a series. When assessing entire proof sequences with neural networks we may be confronted with several thousand steps. Therefore it is interesting to ask whether mappings can be approximated for inputs of arbitrary length or height, *i.e.*, in the maximum norm as well.

In general, this is not possible in very simple situations. The following example is due to [31].

Example 3.3.1. Assume a finite set of continuous activation functions is specified. Assume $\epsilon > 0$. Even for $\Sigma = \{1\}$ a function $f : \Sigma^* \to \mathbb{R}$ exists which cannot be approximated with accuracy ϵ in the maximum norm by a recurrent network with activation functions from the above set regardless of the network's architecture.

Proof. f is constructed by a standard diagonalization argument; in detail, the construction goes as follows: For any number $i \in \mathbb{N}$ there exists only a finite number of recurrent architectures with at most i neurons and activation functions from the above set. For any fixed architecture A and an input $x = [1, \ldots, 1]$ of length i the set $s_A^i = \{y \mid y$ is the output of x via a network of architecture A with weights absolutely bounded by $i\}$ is a compact set. Therefore we can find for any $i \in \mathbb{N}$ a number y_i such that the distance of y_i to $\bigcup_{A \text{ has at most } i \text{ neurons}} s_A^i$ is at least ϵ. The mapping $f : \underbrace{[1, \ldots, 1]}_{i \text{ times}} \mapsto y_i$ cannot be approximated by any recurrent network in the maximum norm with accuracy ϵ and activation function from the specified set.   □

Note that the function f constructed above is computable if an appropriate approximation of the activation functions can be computed. This holds for all standard activations like the sigmoidal function or polynomials. However, the function we have constructed does not seem likely to occur in practical applications. We can prevent the above argumentation from occurring by considering only functions with bounded range. But even here, some functions cannot be approximated, although we have to enlarge the alphabet Σ to a binary alphabet in the case $k = 1$.

Example 3.3.2. Assume an activation function is specified which is for any point Lipschitz continuous in some neighborhood of the point and which has a bounded range. Assume $\epsilon > 0$. Even for $\Sigma = \{0, 1\}$ there exists a function $f : \Sigma^* \to \Sigma$ that cannot be approximated by a recurrent network with the above activation function.

Proof. As already mentioned, it is shown in [115] that any computation with a recurrent network with activation function as specified above can be simulated by a nonuniform Boolean circuit family with polynomially growing resources. An approximation of a mapping between input sequences of a binary alphabet and binary outputs can be seen as a very simple computation in linear time and therefore can be simulated by circuits as well.

It remains to show that some function $f : \Sigma^* \to \Sigma$ exists, which cannot be computed by such a circuit family. It can be shown by a standard counting argument: Any Boolean circuit with n inputs and b gates can be obtained as one of the following circuits: Enumerate the b gates. Choose for any gate one of the 3 activations in $\{AND, OR, NOT\}$ and up to $b - 1 + n$ predecessors of the gate. Choose one of the gates as the output. This leads to at most $p(b, n) = 3^b \cdot (b + n)^{b(b-1+n)} \cdot b$ different circuits. On the contrary, there exist 2^{2^n} functions $\Sigma^n \to \Sigma$. We can find numbers $n_1, n_2, \ldots$ that tend to ∞ and for any i a function $f_i : \Sigma^{n_i} \to \Sigma$ such that $p(n_i^i, n_i) < 2^{2^{n_i}}$ and f_i cannot be implemented by a circuit with n_i^i gates. Any f with $f|\Sigma^{n_i} = f_i$ cannot be implemented by a circuit with polynomial resources and therefore cannot be implemented by a recurrent network as specified above either.   □

In fact, it is not necessary to use the simulation via Boolean circuits. We can argue with networks directly in the same way because the number of mappings a fixed architecture can implement is limited in a polynomial way for common activation functions.

Example 3.3.3. Assume $\Sigma = \{0, 1\}$. Assume an activation function σ is specified such that σ is locally C^1 with $\sigma' \neq 0$. Assume the number of input sequences in Σ^* of length at most T on which any mapping to $\{0, 1\}$ can be approximated in the maximum norm with accuracy < 0.5 by an appropriate choice of the weights in a recurrent architecture with n neurons is limited by a function $p(T, n)$ such that p is polynomial in T. Then there exists a function $f : \Sigma^* \to \Sigma$ which cannot be approximated in the maximum norm by a recurrent network with activation function σ.

Proof. Note that for any finite set of sequences and recurrent networks with at most N neurons we can find a recurrent architecture with N^2 neurons which approximates each single network by an appropriate choice of the weights on the input sequences. This is due to the fact that feed-forward networks with N nodes can be simulated with a standard MLP architecture with N^2 nodes such that the input and output units can be taken as identical for each simulation. Furthermore, the identity can be approximated by $(\sigma(x_0 + \epsilon x) - \sigma(x_0))/(\epsilon \sigma'(x_0))$.

Now we consider input sequences in $\{0, 1\}^*$ and construct outputs such that the number of neurons of a minimal network which approximates the values increases in each step.

Assume that we have constructed the mapping up to sequences of length T_0, and let N_0 be the minimum number of neurons such that a network exists which maps the sequences correctly. There exist 2^T different sequences in $\{0, 1\}^T$. Any network with N_0 neurons can be simulated in a uniform manner by a network with N_0^2 neurons which can approximate any mapping on at most $p(T, N_0^2)$ input sequences of length T. Take T such that $T > T_0$ and $2^T > p(T, N_0^2)$. Then there exists at least one mapping on sequences of length T that cannot be approximated by a network with N_0 neurons.   □

The function f we have constructed is computable if it can be decided whether a given finite set of binary sequences and desired outputs can be mapped correctly by an architecture with activation σ and an appropriate choice of the weights. This is valid for piecewise polynomial activations because the arithmetic of the real numbers is decidable and it is shown for the sigmoidal function as well in [86] modulo the so-called Schanuel conjecture in number theory.

The maximum number of points which can be mapped to arbitrary values approximating $\{0,1\}$ by an architecture is limited by the so-called pseudo-dimension of the architecture. This dimension measures the capacity of the architecture and is limited by a polynomial in the input length for piecewise polynomial functions or the sigmoidal function, respectively [64, 86]. In fact, since it plays a key role when considering the generalization capability of a network we will have a closer look at it in the next chapter.

Of course, the above examples transfer directly to the case $k \geq 2$ because the situation $k = 1$ results when the inputs to a folding network are restricted to trees where any node has at most one nonempty successor. For $k \geq 2$ the situation in the last two examples can be further modified.

Corollary 3.3.1. *Assume the activation is as in one of the previous two examples. Assume $k \geq 2$. Then there exists a mapping $f : \{1\}^*_k \to \{0,1\}$ which cannot be approximated by a folding network in the maximum norm.*

Proof. Each binary input sequence $[a_1, \ldots, a_n]$ can be substituted by a tree with unary labels of the form $1(\tilde{a}_n, 1(\tilde{a}_{n-1}, 1(\ldots, 1(\tilde{a}_1, 1(\bot, \bot)) \ldots)))$, for $k = 2$ where $\tilde{a}_i = \begin{cases} 1(\bot, \bot) & a_i = 1, \\ \bot & a_i = 0. \end{cases}$ $\square$

3.3.2 Approximation on Unary Sequences

Consequently even very restricted and computable mappings exist that cannot be approximated in the maximum norm by a folding architecture. On the contrary, if we consider recurrent networks as a computational model, then the functions which can be computed by a recurrent network contain the functions which can be computed by a Turing machine as a proper subclass [114, 115]. Another demonstration of the computational capability of recurrent networks is the following result:

Theorem 3.3.1. *Assume σ is a continuous squashing function. Then any function $f : \{1\}^* \to [0,1]$ can be approximated in the maximum norm by a recurrent network without hidden layer, activation function σ in the recursive part, and one linear neuron in the feed-forward part. The number of neurons which is sufficient for an approximation can be limited with respect to the accuracy ϵ.*

Proof. Choose $n \in \mathbb{N}$, n even with $1/n \leq \epsilon$. Define $x_i = 1/(2n) + (i-1)/n$ for $i = 1, \ldots, n$, $I_i = [x_i - 1/(4n), x_i + 1/(4n)]$. Since σ is a squashing function we can choose $K > 0$ such that

$$\sigma(Kx) \begin{cases} > 1 - 1/(8n^2) & \text{if } x > 1/(4n), \\ < 1/(8n^2) & \text{if } x < -1/(4n). \end{cases}$$

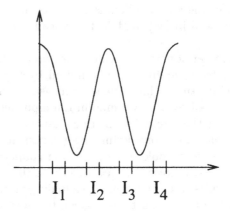

Fig. 3.3. Function g with the property $g(I_j) \supset \cup_{i=1}^4 I_i$ for $j = 1, \ldots, 4$. Such a function leads to a universal approximation property when applied recursively to unary input sequences.

The function

$$g(x) = \sum_{i=1}^n \sigma((-1)^i K \cdot (x - x_i)) - (n/2 - 1)$$

has the property $g(I_j) \supset \cup_{i=1}^n I_i$ for any $j = 1, \ldots, n$ because of the continuity of g and

$$g(x_j - 1/(4n)) \;\leq\; \sum_{i<j, i\,\text{even}} 1 + \sum_{i<j, i\,\text{odd}} 1/(8n^2) + 1/(8n^2)$$
$$+ \sum_{i>j, i\,\text{even}} 1/(8n^2) + \sum_{i>j, i\,\text{odd}} 1 - (n/2 - 1)$$

$$\leq\; (n+1)/(8n^2) \leq 1/(4n)$$

if j is even, $g(x_j - 1/(4n)) \geq 1 - 1/(4n)$ if j is odd, $g(x_j + 1/(4n)) \geq 1 - 1/(4n)$ if j is even, and $g(x_j + 1/(4n)) \leq 1/(4n)$ if j is odd (see Fig. 3.3.2).

g is trivially expanded to inputs from $\mathbb{R}^2$ by just ignoring the first component of the input. The function $\tilde{g}_y : \mathbb{R}^* \to \mathbb{R}$ can be implemented by a recurrent network with n neurons with activation function σ in the recursive function part and a linear output in the feed-forward part. The linearity in g as defined above is considered to be part of the weights in the networks g and h, respectively. Furthermore, $\tilde{g}_y$ can approximate any function $f : \{1\}^* \to [0,1]$ in the maximum norm with accuracy ϵ by an appropriate choice of y. It is sufficient to choose an initial context in $\bigcap_{i\in\mathbb{N}}(g^i)^{-1}(I_{k_i})$ if $f(\underbrace{(1, \ldots, 1)}_{i\,\text{times}}) \in [x_{k_i} - 1/(2n), x_{k_i} + 1/(2n)]$. Note that such a value exists because I_{k_i} is compact, g is continuous, and for any finite number i_0 the intersection $\bigcap_{i\leq i_0}(g^i)^{-1}(I_{k_i})$ is not empty because of the property $g(I_j) \supset \cup_i I_i$. Just start at an arbitrary value in I_{i_0} and successively choose an inverse image in the desired interval I_{i_0-1}, I_{i_0-2}, ... Since in a network implementation the recursive part consists of only one layer with squashing activation, this value y has to be changed to an initial context $(y+n/2-1, 0, \ldots, 0)$ according to the function g. □

The construction can be expanded to show that a sigmoidal network can compute any mapping on off-line inputs in exponential time. An equivalent result is already known in the case of the semilinear activation function [115], but for the standard sigmoidal function only the Turing capability with an exponential increase of time is established in [66]. Since we deal with the sigmoidal function the outputs are only required to approximate the binary values $\{0, 1\}$.

Corollary 3.3.2. *Assume the encoding function* code $: \{0,1\}^+ \to \mathbb{R}$ *is computed by* code$([x_1, \ldots, x_T]) = (3^{-\sum_{i=1}^{T} x_i 2^{i-1}} + 1)/2$. *Then a recurrent architecture exists where only the initial context is not specified with the standard sigmoidal activation function, a fixed number of neurons, and the property: For any possibly partial function* $f : \{0,1\}^+ \to \{0,1\}$ *exists an initial context* y *such that for any input sequence* x *the recurrent network with initial context* $(\text{code}(x), y)$ *computes* $f(x)$ *on off-line inputs.*

Proof. It follows from Theorem 3.3.1 that a recurrent network $h_1 \circ (\tilde{g}_1)_y$ exists with a linear unit h_1, a sigmoidal network g_1, and the property

$$h_1 \circ (\tilde{g}_1)_y(\underbrace{[\top, \ldots, \top]}_{n \text{ places}}) \begin{cases} < 0.1 & \text{if } f(x) = 0, \\ \in \,]0.45, 0.55[& \text{if } f(x) \text{ is not defined}, \\ > 0.9 & \text{if } f(x) = 1, \end{cases}$$

where $n = \sum_i x_i 2^i$ and $x = [x_1, \ldots, x_T]$. Only the initial context y depends on f. The other weights can be chosen as fixed values. Additionally, there exists a network $h_2 \circ (\tilde{g}_2)_{\text{code}(x)}$ with a sigmoidal network g_2 and a linear unit h_2 such that

$$h_2 \circ (\tilde{g}_2)_{\text{code}(x)}(\underbrace{[\top, \ldots, \top]}_{n \text{ places}}) \begin{cases} < 0.15 & \text{if } n < \sum_i x_i 2^i, \\ \in \,]0.2, 0.4[& \text{if } n = \sum_i x_i 2^i, \\ > 0.7 & \text{if } n > \sum_i x_i 2^i \end{cases}$$

with $x = [x_1, \ldots, x_T]$, as we will show later. The simultaneous computation of $h_1 \circ (\tilde{g}_1)_y$ and $h_2 \circ (\tilde{g}_2)_{\text{code}(x)}$ with outputs o_1 and o_2, respectively, is combined with the computation

$$(o_1 > 0.9) \wedge o_2 \in \,]0.2, 0.4[$$

for the output line and

$$((o_1 > 0.9) \vee (o_1 < 0.1)) \wedge o_2 \in \,]0.2, 0.4[$$

for the line indicating whether output is present. This latter computation can be approximated arbitrarily well in the feed-forward part of a sigmoidal network because the identity, the perceptron activation, and Boolean connections can be approximated. Therefore, the entire construction leads to a sigmoidal architecture that computes f on off-line inputs. Only the part $\mathbf{y}$ of the initial context depends on f.

It remains to show that a mapping $h_2 \circ \tilde{g}_2$ with the demanded properties exists. The recursive mapping induced by $x \mapsto 3x$ computes on the initial context $3^{-\sum_i x_i 2^i - 1}$ the value $3^{-\sum_i x_i 2^i - 1 + n}$ for an input of length n. This mapping is combined with the function tanh which approximates the identity for small values. In fact, the function induced by $g(x) = \tanh(3x)$ computes on the initial context $3^{-\sum x_i 2^i - 1}$ and inputs of length n a value y with

$$
y \begin{cases} > 0.7 & \text{if } n \geq \sum x_i 2^i + 1 \\ \in \ \big](1 - 3^{2(-\sum x_i 2^i - 1 + n)})3^{-\sum x_i 2^i - 1 + n}, \\ \qquad (1 + 3^{2(-\sum x_i 2^i - 1 + n)})3^{-\sum x_i 2^i - 1 + n}\big[& \text{otherwise}. \end{cases}
$$

Since $|1 - \tanh(x)/x| < x^2/3$ for $x \neq 0$ this can be seen by induction: If $\tilde{g}_{3-k}(\mathsf{T}^n) \in \ ](1 - 3^{2(-k+n)})3^{-k+n}, (1 + 3^{2(-k+n)})3^{-k+n}[$ for $k > n+1$, then

$$
\begin{aligned}
& \tilde{g}_{3-k}(\mathsf{T}^{n+1}) \\
=\ & \tanh(3 \cdot \tilde{g}_{3-k}(\mathsf{T}^n)) \\
\in\ & \big]3\,\tilde{g}_{3-k}(\mathsf{T}^n)\left(1 - 3\,\tilde{g}_{3-k}(\mathsf{T}^n)^2\right), 3\,\tilde{g}_{3-k}(\mathsf{T}^n)\left(1 + 3\,\tilde{g}_{3-k}(\mathsf{T}^n)^2\right)\big[\\
\subset\ & \big]3^{-k+n+1}\left(1 - 3^{2(-k+n)}\right)\left(1 - 3\,(3^{-k+n})^2(1 - 3^{2(-k+n)})^2\right), \\
& \quad 3^{-k+n+1}\left(1 + 3^{2(-k+n)}\right)\left(1 + 3\,(3^{-k+n+1})^2(1 + 3^{2(-k+n)})^2\right)\big[\\
\subset\ & \big]3^{-k+n+1}\left(1 - 3^{2(-k+n+1)}\right), 3^{-k+n+1}\left(1 + 3^{2(-k+n+1)}\right)\big[.
\end{aligned}
$$

For $k \leq n+1$ we obtain

$$
\tilde{g}_{3-k}(\mathsf{T}^{n+1}) \geq \tanh(3(3^{-1}(1 - 3^{-2}))) = \tanh(8/9) > 0.7 .
$$

Here T^n denotes the sequence of length n with elements T. The function $h_2 \circ \tilde{g}_2$ with the desired properties can be obtained by exchanging $\tanh(x)$ with $2 \cdot \mathrm{sgd}(2x) - 1$. $\qquad\square$

Consequently even very small recurrent networks with sigmoidal activation function can compute any function in exponential time. The proof can be transferred to more general squashing functions.

Theorem 3.3.2. *Assume σ is a continuous squashing function which is three times continuously differentiable in the neighborhood of at least one point x_0 such that $\sigma'(x_0) \neq 0$, $\sigma''(x_0) = 0$. Then a recurrent architecture $h \circ \tilde{g}_-$ with unspecified initial context exists such that for every possibly partial function*

$f : \{0,1\}^{+} \rightarrow \{0,1\}$ *some* y *can be found such that* $h \circ \tilde{g}_{(-,y)}$ *computes* f *on off-line inputs. The encoding function is* $\mathrm{code}([x_1,\ldots,x_T]) = (3^{-\sum_{i=1}^{T} x_i 2^i - 1} + 1)/2$.

Proof. Because of Theorem 3.3.1 a recurrent network $h_1 \circ (\tilde{g}_1)_y$ exists which outputs a value smaller than 0.1 for inputs of length n if $f(x)$ is 0, x being the sequence corresponding to the number n, which outputs a value at least 0.9 if $f(x)$ is 1, and a value in the interval $]0.45, 0.55[$ if $f(x)$ is not defined. Additionally, a network which outputs a value of at most 0.15, or in $]0.2, 0.4[$, or at least 0.7 for an input of length shorter than n, equal to n, or longer than n, respectively, exists as we will show later. Hence the combination of the two networks with the function $(o_1 > 0.9) \wedge o_2 \in]0.2, 0.4[$ for the output line and $((o_1 > 0.9) \vee (o_1 < 0.1)) \wedge o_2 \in]0.2, 0.4[$ for the line indicating whether output is present, o_1 and o_2 denoting the outputs of the two recurrent networks, yields the desired result. Since the identity and perceptron activation can be approximated, a network with activation function σ results.

In order to construct g_2 choose ϵ such that $|\epsilon^2 \sigma'''(x)/\sigma'(x_0)| < 2$ for all $|x - x_0| < \epsilon$. The function

$$\bar{\sigma}(x) = \frac{\sigma(\epsilon x + x_0) - \sigma(x_0)}{\sigma'(x_0)\epsilon}$$

fulfills the property $|1 - \bar{\sigma}(x)/x| < x^2/3$ for every $x \neq 0$, $|x| < 1$ because

$$\bar{\sigma}(x) = x + \frac{\sigma'''(\bar{x})\epsilon^2 x^3}{6\sigma'(x_0)}$$

for $|x| < 1$ and some point $\bar{x}$ between x_0 and $x_0 + \epsilon x$ and hence

$$\left|1 - \frac{\bar{\sigma}(x)}{x}\right| = \left|\frac{\sigma'''(\bar{x})\epsilon^2}{6\sigma'(x_0)}\right| x^2 < \frac{x^2}{3}.$$

Hence the function induced by $g(x) = \bar{\sigma}(3x)$ computes on the initial context $3^{-\sum x_i 2^i - 1}$ and inputs of length n a value y with

$$y \begin{cases} > 0.7 & \text{if } n \geq \sum x_i 2^i + 1 \\ \in \left](1 - 3^{2(-\sum x_i 2^i - 1 + n)})3^{-\sum x_i 2^i - 1 + n}, \right. \\ \quad \left.(1 + 3^{2(-\sum x_i 2^i - 1 + n)})3^{-\sum x_i 2^i - 1 + n}\right[& \text{otherwise.} \end{cases}$$

as can be seen by induction. The affine mapping in g can be integrated in the weights of the recursive mapping. Hence a function g_2 with the desired properties results. $\qquad \square$

However, the proof mainly relies on the fact that the amount of data that can be stored in the internal states is unlimited and the computation is performed with perfect accuracy. If the internal stack is limited, e.g., because the activation function has a finite range like the perceptron activation function, it can be seen immediately that the computational power is restricted to finite automata. Furthermore, the number of neurons necessarily increases if the function to be computed becomes more complex.

3.3.3 Noisy Computation

Even in the case of a real valued smooth activation function the accuracy of the computation may be limited because the input data is noisy or the operations on the real numbers are not performed with perfect reliability. To address this situation, computations that are subject to some limited smooth noise are considered in [84]. It turns out that the capability of recurrent networks also reduces to finite automata for sigmoidal activation functions. The argumentation transfers directly to folding networks, indicating that in practical applications they will only approximate mappings produced by Mealy tree automata correctly when dealing with very large trees.

In order to obtain explicit bounds on the number of states of a tree automaton which is equivalent to a noisy folding network we briefly carry out a modification of [84] for folding networks. We restrict the argumentation to the following situation:

We consider a folding network $h \circ \tilde{g}_y$, where h and g are measurable, which computes a $\{0, 1\}$-valued function on $(\mathbb{R}^m)_k^*$. We can drop the feed-forward part h by assuming that a tree is mapped to 1 if and only if the image under $\tilde{g}_y$ lies in a distinguished set of final states F, the states which are mapped to 1 using h. Now the computation $\tilde{g}_y$ is affected with noise, *i.e.*, it is induced by a network function $g : \mathbb{R}^{m+k \cdot n} \to \mathbb{R}^n$ composed with a noise process $Z : \mathbb{R}^n \times \mathcal{B}(\mathbb{R}^n) \to [0, 1]$ such that $Z(q, A)$ describes the probability that a state q is changed into some state contained in A by some noise. $\mathcal{B}(\mathbb{R}^m)$ are the Borel measurable sets in $\mathbb{R}^n$.

Assume that the internal states are contained in a compact set $\Omega \subset \mathbb{R}^n$, *e.g.*, $[0, 1]^n$; assume that Z can be computed by $Z(q, A) = \int_{q' \in A} z(q, q') d\mu$ for some measure μ on Ω and a measurable function $z : \Omega^2 \to [0, \infty[$. The probability that one computation step in $\tilde{g}_y$ on the initial states $(q_1, \ldots, q_k)$ and input a results in a state q' can be described by

$$\pi_a(q_1, \ldots, q_k, q') = z(g(a, q_1, \ldots, q_k), q').$$

The probability that a computation on an input tree t with root a and subtrees t_1, ..., t_k transfers initial states $\mathbf{q}^1$, ..., $\mathbf{q}^k$, where the dimensions depend on the structure of the t_i, to a state q' can be recursively computed by $\pi_t(\mathbf{q}^1, \ldots, \mathbf{q}^k, q') =$

$$\int_{(q_1'', \ldots, q_k'') \in \Omega^k} \pi_{t_1}(\mathbf{q}^1, q_1'') \cdot \ldots \cdot \pi_{t_k}(\mathbf{q}^k, q_k'') \cdot \pi_a(q_1'', \ldots, q_k'', q') d\mu^k.$$

The probability that an entire computation on a tree t starting with the initial context $\mathbf{y}$ leads to a value 1 is described by the term $\int_{q \in F} \pi_t(\mathbf{y}, \ldots, \mathbf{y}, q) d\mu$. A tree is mapped to 1 by a *noisy computation* if this probability is at least $0.5 + \rho$ for a fixed positive value ρ. It is mapped to 0 if this probability is at most $0.5 - \rho$. We refer to a noisy computation with $\bar{g}_{\mathbf{y}}^{\text{noise}}$.

A final assumption is the fact that the noise process is piecewise equicontinuous, *i.e.*, one can find a division of Ω into parts $\Omega_1, \ldots, \Omega_l$ such that

$$\forall \epsilon > 0 \, \exists \delta > 0 \, \forall p \in \Omega \, \forall q', q'' \in \Omega_j \, (|q' - q''| < \delta \Rightarrow |z(p, q') - z(p, q'')| < \epsilon).$$

This situation is fulfilled, for example, if the state q is affected with some clipped Gaussian noise.

Under these assumptions, a folding network behaves like a tree automaton. Here a *tree automaton* is a tuple (I, S, δ, s, F), where I is a set, the input alphabet, S is a finite set with $S \cap I = \emptyset$, the set of internal states, $\delta : I \times S^k \to S$ is the transition function, $s \in S$ is the initial state, and $F \subset S$ is a nonempty set of final states. A tree automaton computes the mapping $f : I_k^* \to \{0, 1\}$, $f(t) = 1 \iff \bar{\delta}_s(t) \in F$.

Theorem 3.3.3. *Assume that under the above conditions a noisy computation $\bar{g}_{\mathbf{y}}^{\text{noise}}$ computes a mapping $(\mathbb{R}^m)_k^* \supset B \to \{0, 1\}$. Then the same mapping can be computed by a tree automaton (I, S, δ, s, F).*

Proof. Define the equivalence relation $t_1 \equiv t_2$ for trees t_1 and t_2 if and only if for all trees t_1' and t_2' the following holds:

Assume t_1' is a tree with subtree t_1, t_2' is the same tree except for t_1 which is substituted by t_2. Then $\bar{g}_{\mathbf{y}}^{\text{noise}}(t_1') = 1 \Leftrightarrow \bar{g}_{\mathbf{y}}^{\text{noise}}(t_2') = 1$.

Assume that only a finite number of equivalence classes $[t]$ exists for this relation. Then $\bar{g}_{\mathbf{y}}^{\text{noise}}$ can be computed by the tree automaton with states $\{[t] \mid t \text{ is a tree}\}$, the initial state $[\bot]$, the final set $\{[t] \mid \bar{g}_{\mathbf{y}}^{\text{noise}}(t) = 1\}$, and the transition function which maps the states $[t_1], \ldots, [t_k]$ under an input a to $[a(t_1, \ldots, t_k)]$. Therefore it remains to show that only a finite number of equivalence classes exists.

Assume t_1 and t_2 are trees such that

$$\int_{q \in \Omega} |\pi_{t_1}(\mathbf{y}, \ldots, \mathbf{y}, q) - \pi_{t_2}(\mathbf{y}, \ldots, \mathbf{y}, q)| d\mu \leq \rho.$$

Then t_1 and t_2 are equivalent. Otherwise, this fact would yield a contradiction because it can be computed for some trees t_1^0 and t_2^0, respectively, which differ only in one subtree equal to t_1 or t_2, respectively:
(We assume w.l.o.g. that the differing subtrees of t_1^0 and t_2^0 are in each layer the leftmost subtrees, *i.e.*, $t_1^i = a_{i+1}(t_1^{i+1}, s_2^{i+1}, \ldots, s_k^{i+1})$ and $t_2^i = a_{i+1}(t_2^{i+1}, s_2^{i+1}, \ldots, s_k^{i+1})$ for $i = 0, \ldots, l - 1$ and some $l \geq 1$, labels a_{i+1} and trees s_j^{i+1}, t_1^{i+1}, t_2^{i+1} with $t_1^l = t_1$ and $t_2^l = t_2$.)

$$
\begin{aligned}
2\rho \;\leq\; & \left| \int_{q \in F} \pi_{t_1^0}(\mathbf{y},\ldots,\mathbf{y},q)d\mu - \int_{q \in F} \pi_{t_2^0}(\mathbf{y},\ldots,\mathbf{y},q)d\mu \right| \\
=\; & \left| \int_{q \in F} \int_{(q_1^1,\ldots,q_k^1) \in \Omega^k} (\pi_{t_1^1}(\mathbf{y},\ldots,\mathbf{y},q_1^1) - \pi_{t_2^1}(\mathbf{y},\ldots,\mathbf{y},q_1^1)) \cdot \right. \\
& \left. \pi_{s_2^1}(\mathbf{y},\ldots,\mathbf{y},q_2^1)\cdot\ldots\cdot\pi_{s_k^1}(\mathbf{y},\ldots,\mathbf{y},q_k^1)\cdot\pi_{a_1}(q_1^1,\ldots,q_k^1,q)d\mu^{k+1} \right| \\
=\; & \ldots \\
=\; & \left| \int_{q \in F} \int_{\mathbf{q} \in \Omega^{lk}} (\pi_{t_1}(\mathbf{y},\ldots,\mathbf{y},q_1^l) - \pi_{t_2}(\mathbf{y},\ldots,\mathbf{y},q_1^l)) \cdot \right. \\
& \pi_{s_2^1}(\mathbf{y},\ldots,\mathbf{y},q_2^1)\cdot\ldots\cdot\pi_{s_k^1}(\mathbf{y},\ldots,\mathbf{y},q_k^1)\cdot\ldots\cdot \\
& \pi_{s_2^l}(\mathbf{y},\ldots,\mathbf{y},q_2^l)\cdot\ldots\cdot\pi_{s_k^l}(\mathbf{y},\ldots,\mathbf{y},q_k^l)\cdot \\
& \pi_{a_1}(q_1^1,\ldots,q_k^1,q)\cdot \\
& \left. \pi_{a_2}(q_1^2,\ldots,q_k^2,q_1^1)\cdot\ldots\cdot\pi_{a_l}(q_1^l,\ldots,q_k^l,q_1^{l-1})d\mu^{lk+1} \right| \\
\leq\; & \int_{q \in \Omega} |\pi_{t_1}(\mathbf{y},\ldots,\mathbf{y},q) - \pi_{t_2}(\mathbf{y},\ldots,\mathbf{y},q)|d\mu \leq \rho \,.
\end{aligned}
$$

$\pi_t(\mathbf{y},\ldots,\mathbf{y},\_)$ is piecewise uniformly continuous for any $t \in B$ with the same constants as z because

$$
\begin{aligned}
& |\pi_t(\mathbf{y},\ldots,\mathbf{y},p) - \pi_t(\mathbf{y},\ldots,\mathbf{y},q)| \\
\leq\; & \int_{\mathbf{p} \in \Omega^k} \pi_{t_1}(\mathbf{y},\ldots,p_1)\cdot\ldots\cdot\pi_{t_k}(\mathbf{y},\ldots,p_k) \\
& \cdot|\pi_a(p_1,\ldots,p_k,p) - \pi_a(p_1,\ldots,p_k,q)|d\mu^{k+1} \\
\leq\; & \epsilon
\end{aligned}
$$

for $|p - q| < \delta$, $p,q \in \Omega_i$, $t = a(t_1,\ldots,t_k)$.

But only a finite number of functions exists which are uniformly continuous on Ω_i with constant δ corresponding to $\epsilon = \rho/(4\mu(\Omega))$ in the assumption made on z, and which have a distance at least ρ from each other with respect to $d\mu$. The number can be bounded explicitly using the same argumentation as in [84], Theorem 3.1: The Ω_i are covered with a lattice of points of distance at most δ, and the values of these points are contained in one of a finite set of intervals with diameter at most $\rho/(2\mu(\Omega))$ which cover $[0,1]$. An upper bound $\Pi_i(2\mu(\Omega)/\rho)^{(\sqrt{n}\cdot\text{diam}(\Omega_i)/\delta)^n}$ results where $\text{diam}(\Omega_i)$ is the diameter of the component Ω_i of Ω. $\qquad\square$

Consequently, the computational capability of folding networks reduces to tree automata if the computation is subject to piecewise equicontinuous noise. The argumentation can be expanded to a function with outputs in a finite, but not necessarily binary set. Then it follows that at most Mealy tree automata can be computed by a noisy folding network.

3.3.4 Approximation on a Finite Time Interval

So far, we have only considered mappings with discrete inputs which we wanted to approximate in the maximum norm. All negative results transfer to the case of continuous labels, of course. But here, an additional question arises: Can any continuous mapping be approximated in the maximum norm with a folding network on restricted inputs? Note that approximation in the maximum norm on restricted inputs is a special case of approximation in probability if we only consider symbolic, *i.e.*, discrete domains. For continuous labels, the following result shows that an approximation is possible. However, the encoding dimension necessarily increases for realistic networks and some functions that are to be approximated for increasing input height.

Theorem 3.3.4. *Choose $T \in \mathbb{N}$ and a compact set $B \subset \mathbb{R}^m$. For any continuous mapping $f : B_k^{\leq T} \to \mathbb{R}^n$ and $\epsilon > 0$ a folding network $h \circ \tilde{g}_\mathbf{y}$ exists such that $|h \circ \tilde{g}_\mathbf{y}(t) - f(t)| < \epsilon$ for all $t \in B_k^{\leq T}$.*

g can be a feed-forward network without hidden layer and h a single hidden layer feed-forward network with linear outputs. The other activation functions are locally Riemann integrable and nonpolynomial or squashing.

If g is continuous and the interior of B is not empty, the encoding dimension increases at least exponentially with T for $k \geq 2$ and linearly with T for $k = 1$ for $\epsilon < 1$ and some real valued and continuous f, regardless of the number of hidden layers in g.

Proof. Without a restriction on the encoding dimension it is easy to construct an encoding g such that $\tilde{g}_\mathbf{y}$ simply writes the single labels of an input tree of height T in one real vector of dimension $d = k^T(m+1)+1$ or $d = T(m+1)+1$ for $k = 1$, respectively. If the actual number of places which are already used, p, is encoded into the last dimension as $(0.1)^p$, and s is a real number which is not contained in any label in B, an encoding is induced by

$$h(\mathbf{x}, \mathbf{x}^1, (0.1)^{l_1}, \ldots, \mathbf{x}^k, (0.1)^{l_k}) =$$
$$(s, \mathbf{x}, x_1^1, \ldots, x_{l_1}^1, \ldots, x_1^k, \ldots, x_{l_k}^k, (0.1)^{l_1+\cdots+l_k+m+1}),$$

where the coefficients $x_1^1, \ldots$ can be computed with a finite gain using $l_1, l_2,$ $\ldots$ Since the finite gain can be approximated, for example, with a sigmoidal network, the entire mapping g can be approximated with a single hidden layer network with linear outputs. The linearity can be integrated into part h and the connections in g. h can be chosen such that it approximates the continuous mapping on these codes in $\mathbb{R}^d$ to $\mathbb{R}^n$ [58].

Surprisingly, this brute force method is in some way the best possible encoding. Let an encoding dimension $l(T)$ be given which does not increase exponentially or linearly if $k = 1$, respectively, with T. Assume g is continuous. Choose $\mathbf{x} \in B$ and $\epsilon > 0$ such that the ball of radius ϵ with center $\mathbf{x}$ is contained in B. Choose T with $md_0 > l(T)k$, where $d_0 = k^{T-1}$ or $T - 1$ if $k = 1$, respectively. Assume $\mathbf{a}^{11}, \ldots, \mathbf{a}^{md_0}$ are different points in B. We consider the following mapping f with images in $[-1, 1]$:

We scale all coefficients of the leaves with a factor such that one value becomes ± 1. Then we output one coefficient depending on the label of the root. Formally, the image of a tree with height T, root $\mathbf{a}^{ij}$ and leaves $\mathbf{a}^1, \ldots,$ $\mathbf{a}^{d_0}$, is $(a_i^j - x_i)/\max\{|a_k^l - x_k| \,|\, k, l\}$ if the point $(\mathbf{a}^1, \ldots, \mathbf{a}^{d_0})$ is not contained in the ball of radius $\epsilon/2$ and center $(\mathbf{x}, \ldots, \mathbf{x})$ in B^{d_0}. Otherwise, f is an arbitrary continuation of this function. If $k = 1$, a mapping f is considered which maps the sequences of length T of the form $[\mathbf{a}^1, \ldots, \mathbf{a}^{d_0}, \mathbf{a}^{ij}]$ to an output $(a_i^j - x_i)/\max\{|a_k^l - x_k| \,|\, k, l\}$ or the value of a continuous completion, respectively, in the same manner.

The approximation $h \circ \tilde{g}_y$ on these trees of height T with variable leaves and root and fixed interior labels or on sequences of length T, respectively, decomposes into a mapping $\bar{g} : B^{d_0} \to \mathbb{R}^{l(T)k}$ and $h \circ g : B \times \mathbb{R}^{l(T)k} \to \mathbb{R}$, where necessarily antipodal points in the sphere of radius ϵ and center $(\mathbf{x}, \ldots, \mathbf{x})$ in B^{d_0} exist which are mapped by $\bar{g}$ to the same value because of the Theorem of Borsuk-Ullam [1]. Consequently, at least one value of $h \circ \tilde{g}_y$ differs from the desired output by at least 1.   □

As a consequence, continuous mappings cannot be approximated in the maximum norm with limited resources for restricted inputs. The fact that it is necessary to increase the encoding dimension is another proof which shows that arbitrary mappings cannot be approximated in the maximum norm on trees with unlimited height with a folding network.

3.4 Discussion and Open Questions

We have shown in the first part of this chapter that folding networks are universal approximators in the sense that any measurable function on trees, even with real labels, to a real vector space can be approximated arbitrarily well in probability. One can derive explicit bounds on the number of layers and neurons that are required for such an approximation. In particular, only a finite number of neurons is needed in the recursive part if the labels are elements of a finite alphabet. This situation takes place in symbolic domains, for example. Furthermore, only a finite number of neurons is necessary in the entire network in order to interpolate a finite set of data exactly; this situation occurs whenever an empirical pattern set is dealt with in a concrete learning scenario.

The main relevance of these results for practical applications is twofold: First, the in principle approximation capability of folding networks is a necessary condition for any practical application where an unknown function is to be learned. If this capability was limited, no training algorithm, however complicated, would succeed in approximation tasks dealing with functions that cannot be represented by folding networks. Second, the explicit bounds on the number of layers and neurons enable us to limit the search space for an appropriate architecture in a concrete algorithm. Furthermore: The explicit bounds on the number of neurons necessary to interpolate a finite set of points make it possible to create test situations for the in principle capability of a concrete learning algorithm of minimizing the empirical error.

We have used an encoding method for symbolic data and, additionally, a discretization process when dealing with continuous data in the proofs. The discretization process controls the number of neurons which is necessary in the recursive part of a network and which is *a priori* unlimited for an arbitrary function. A consideration of the smoothness, *i.e.*, the derivatives of the mapping that has to be approximated allows us to limit the number of neurons that are necessary in the recursive part. This is due to the fact that a bound on the derivative also limits the constants that appear in a continuity requirement. Therefore we can limit the diameter of the intervals the input space has to be divided into *a priori*. Unfortunately, the bounds on the number of neurons that are obtained in this way are rather large compared to the case of finite inputs. Can they be improved in the case of real labels, too?

All approximation results in the first part lead to approximation in probability. If a continuous function is to be approximated in the maximum norm for restricted inputs, the encoding is necessarily a trivial encoding in the worst case. This fact indicates that an approximation of functions that are very sensitive to small changes of the input labels may lead to problems in practical application. Considering approximation in the maximum norm for purely symbolic inputs touches on computability questions. It has been shown in the second part of this chapter that functions exist that cannot be approximated in the maximum norm with a recurrent network. Indeed, under realistic assumptions, *i.e.*, the presence of noise, only Mealy tree automata can be approximated. As a practical consequence the capability of using recurrent networks in applications dealing with very large trees or very long sequences is restricted.

However, it is an interesting fact that from a theoretical point of view even noncomputable functions can be computed by a recurrent network if the computation on the real values is performed with perfect reliability. In fact, we have shown that any function can be computed with a sigmoidal recurrent network and a fixed number of neurons in exponential time on off-line inputs. But the question is still open as to whether it is possible to simulate simple Turing machines with recurrent networks with the standard sigmoidal activation function and only a polynomial increase in time.

Chapter 4

Learnability

Frequently, a finite set of training examples $(x_1, y_1), \ldots, (x_m, y_m)$ is available in a concrete learning task. We choose a folding architecture with an appropriate number of layers and neurons – knowing that the architecture can represent the data perfectly and approximate the regularity in accordance to which the inputs x_i are mapped to the y_i in principle. Then we start to fit the parameters of the architecture. Assumed that we manage the task of minimizing the error on the given data, we hope that the network we have found represents both the empirical data and the entire mapping correctly. That means that the network's output should be correct for even those data x which are different from the x_i used for training.

Here the question arises as to whether there exists any guarantee that a network generalizes well to unseen data when it has been trained such that it has small empirical error on the training set. That is, the question arises as to whether a finite set of empirical data includes sufficient information such that the underlying regularity can be learned with a folding architecture. Again, a positive answer to this question is a necessary condition for the success of any learning algorithm for folding networks. A negative answer would yield the consequence that a trained network may remember the empirical data perfectly – but no better than a table-lookup, and the network may have an unpredictable behavior with unseen data.

Of course the question of learnability of a function class from an information theoretical point of view can be made more precise in many different ways dealing with different scenarios. Here we consider the classical PAC setting proposed by Valiant [129] and some of its variants: We assume that the empirical data is given according to an – in general unknown – probability distribution. The underlying function can be learned if we can bound the probability of poor generalization of our learning algorithm in dependence on the number of examples, the number of parameters in our network, and maybe some other quantities that occur in the algorithm. We require that the probability of training samples where our learning algorithm outputs a function that differs significantly from the underlying regularity becomes small if the number of examples increases. Explicit bounds on the number of examples necessary for valid generalization, such that we can limit the amount of data *a priori* that we need, are of special interest. Since the computing time of a learning algorithm directly depends on the number of examples that have to be considered, these bounds should be polynomial in the various parameters of the training algorithm.

Even the demand for a probably good generalization with an increasing number of examples can be specified in several ways:

Usually we deal with an unknown but fixed distribution of the data such that learnability with respect to one fixed distribution is sufficient. Maybe we are aware of additional information about the data and can restrict the consideration to a special class of distributions. We can, for example, limit the maximum height of input trees if large trees rarely occur. On the other hand, we may be interested in distribution-independent bounds because we do not know the special circumstances of our learning scenario. But we only want to train the network once with a number of examples that is sufficient for every scenario.

Of course one algorithm that generalizes well is sufficient. Therefore we may ask whether at least one good algorithm exists. But although such an algorithm might exist this fact could be useless for us because this training method has high computational costs. On the other hand, we may ask for a guarantee that any algorithm that produces small empirical error or satisfies other properties generalizes well. Such a result allows us to improve the performance of our training method without losing the generalization capability – as long as we take care that the algorithm produces small empirical error.

Due to the computational costs, noisy data, or some other reasons perfect minimization of the empirical error may be impossible. In these cases a result in which the empirical error is representative for the real error would be interesting.

Finally, we can specify the learning scenario itself in different ways: We may be interested in the classification of data or in the interpolation of an entire real function. The function that is to be learned may be of the same form as the network architecture we want to fit to the data or it may have an entirely different and unknown form. Moreover, the data may be subject to some noise such that we have to learn a probability distribution instead of a simple function.

The above questions arise in all these settings :

1. Does a learning algorithm exist that generalizes well?
2. Can the generalization ability of every algorithm with small empirical error be established?
3. In which cases is the empirical error representative of the real error?
4. What amount of data is necessary in the above questions? Do explicit, preferably distribution-independent bounds exist?

We start this chapter with a formal definition of the learning scenario and an overview about well known results which will be used later on. Here we mainly use the notation of [132] and refer to the references therein. Afterwards we have a closer look at the distribution-dependent setting and add some general results to this topic. The so-called VC-dimension and generalizations, respectively, play a key role for concrete bounds. The VC-dimension of folding networks depends on several network parameters and, in the most interesting cases, even on the maximum height of the input trees. Consequently, explicit bounds on the number of examples can be derived in all of the above questions for the distribution-dependent setting. Distribution-independent bounds cannot exist in general. Examples can be constructed in which the amount of data increases exponentially compared to the learning parameters, due to the considered distribution. But even without any explicit prior knowledge about the distribution, valid generalization can be guaranteed in some way – in particular, when the data turns out to behave well a posteriori, that is, if the height of the trees in the training set is limited. This consideration fits into the framework of so-called luckiness functions.

4.1 The Learning Scenario

Learning deals with the possibility of learning an abstract regularity given a finite set of data. We fix an input space X (for example, the set of lists or trees) which is equipped with a σ-algebra. We fix a set $\mathcal{F}$ of functions from X to $[0, 1]$ (a network architecture, for example). An unknown function $f : X \to [0, 1]$ is to be learned with $\mathcal{F}$. For this purpose a finite set of independent, identically distributed data $\mathbf{x} = (x_1, \ldots, x_m)$ is drawn according to a probability distribution P on X.

A *learning algorithm* is a mapping

$$h : \bigcup_{m=1}^{\infty} (X \times Y)^m \to \mathcal{F}$$

which selects a function in $\mathcal{F}$ for any pattern set such that this function – hopefully – nearly coincides with the function that is to be learned. We write $h_m(f, \mathbf{x})$ for $h_m(x_1, f(x_1), \ldots, x_m, f(x_m))$. Since we are only interested in the information theoretical part in this chapter and will consider computational issues in the next chapter, an algorithm reduces to a mapping as defined above without taking care about whether and how it can be implemented. An algorithm tries to output a function of which it knows only a number of training examples.

If the algorithm yields valid generalization, the *real error* $d_P(f, h_m(f, \mathbf{x}))$ where

$$d_P(f, g) = \int_X |f(x) - g(x)| dP(x)$$

will highly probably be small. Note that d_P defines a pseudometric on the class of measurable functions. Of course, this error is unknown in general since the probability P and the function f that has to be learned are unknown.

A concrete learning algorithm often simply minimizes the *empirical error*
$\hat{d}_m(f, h_m(f, \mathbf{x}), \mathbf{x})$ where

$$\hat{d}_m(f, g, \mathbf{x}) = \frac{1}{m} \sum_{i=1}^{m} |f(x_i) - g(x_i)| .$$

For example, a standard training algorithm for a network architecture fits
the weights by means of a gradient descent on the surface representing the
empirical error in dependence on the weights.

4.1.1 Distribution-dependent, Model-dependent Learning

In the *distribution-dependent setting* we consider only one fixed probability
P on X. We first assume that the function f that is to be learned is itself
contained in $\mathcal{F}$. Hence the class $\mathcal{F}$ is a model for f. Since $f \in \mathcal{F}$, a concrete
algorithm can produce empirical error 0 for every function f and training
sample $\mathbf{x}$; such an algorithm is called *consistent*. A class with outputs in
the finite set $\{0, 1\}$ is called a *concept class*. A concept has an extremely
simple form, corresponding to a classification into two sets. A generalization
of concepts are functions with values in a finite alphabet which correspond
to a classification of the space X into a finite set of classes.

We denote by P^m the product measure induced by P on X^m. Both here
and in the following, all functions and sets we consider have to be measurable.
See, *e.g.*, [4, 51] for conditions that guarantee this property.

Definition 4.1.1. *A function set $\mathcal{F}$ is said to be* probably approximately
correct *or* PAC learnable *if an algorithm h exists (which is then called PAC,
too) such that for any $\epsilon > 0$ and $\delta > 0$ a number $m_0 \in \mathbb{N}$ exists such that for
every $m \geq m_0$*

$$\sup_{f \in \mathcal{F}} P^m(\mathbf{x} \mid d_P(f, h_m(f, \mathbf{x})) > \epsilon) \leq \delta .$$

$\mathcal{F}$ is called probably uniformly approximately correct *or* PUAC learnable *if
an algorithm h exists (which is then called PUAC, too) such that for any
$\epsilon > 0$ and $\delta > 0$ a natural number m_0 exists such that for every $m \geq m_0$*

$$P^m(\mathbf{x} \mid \sup_{f \in \mathcal{F}} d_P(f, h_m(f, \mathbf{x})) > \epsilon) \leq \delta .$$

$\mathcal{F}$ is said to be consistently PAC learnable *if any consistent algorithm is PAC.
$\mathcal{F}$ is said to be* consistently PUAC learnable *if any consistent algorithm is
PUAC.*

$\mathcal{F}$ has the property of uniform convergence of empirical distances *or*
UCED *for short if for any $\epsilon > 0$ and $\delta > 0$ a number $m_0 \in \mathbb{N}$ exists such
that for every $m \geq m_0$*

$$P^m(\mathbf{x} \mid \exists f, g \in \mathcal{F} | \hat{d}_m(f, g, \mathbf{x}) - d_P(f, g) | > \epsilon) \leq \delta .$$

If a fixed ϵ is considered this parameter is referred to as accuracy. *If a
fixed δ is considered this δ is called* confidence.

PAC learnability can be seen as the weakest condition for efficient learnability. The number of examples that are necessary for a probably good output of an algorithm is limited independently of the (unknown) function that
has to be learned. If such a uniform bound does not exist, the amount of
data necessary for valid generalization cannot be determined *a priori*. In this
case, a correct output takes more examples and consequently more time than
expected for at least some situations. Of course, efficient learnability requires
additionally that the number of examples is only polynomial in the required
accuracy and confidence and that the algorithm runs in polynomial time in
dependence on the data. In the original work of Valiant the question of the
complexity of learning is included in the PAC framework [129]. In this chapter we only consider the information theoretical point of learning, and ask
the question of computational complexity in the next chapter.

Obviously, the UCED property implies consistent PUAC learnability,
which itself implies consistent PAC learnability. But the UCED property
is a stronger condition than consistent PUAC learnability (see [132] Example 6.4). As already stated, consistent learnability is desirable such that we
can use any efficient algorithm with minimum empirical error. The UCED
property is interesting if we do not manage to minimize the empirical error
due to high computational costs, because this property leads to the fact that
the empirical error of the training algorithm is representative for the real
error. Finally, the UCED property does not refer to the notion of a learning
algorithm.

It would be desirable to obtain equivalent characterizations for the other
terms, too, such that the characterizations do not refer to the notion of a
learning algorithm and can be tested more easily if only $\mathcal{F}$ is known. For this
purpose several quantities are introduced.

Definition 4.1.2. *For a set S with pseudometric d the* covering number
$N(\epsilon, S, d)$ *denotes the smallest number n such that n points $x_1, \ldots, x_n$ in S
exist such that the closed balls with respect to d with radius ϵ and center x_i
cover S.*

The packing number $M(\epsilon, S, d)$ *is the largest number n such that n points
$x_1, \ldots, x_n$ in S exist that are ϵ-separated, that is, $d(x_i, x_j) > \epsilon$ for $i \neq j$.*

It follows immediately that $M(2\epsilon, X, d) \leq N(\epsilon, X, d) \leq M(\epsilon, X, d)$. Learnability can be characterized in terms of the covering number of $\mathcal{F}$ as follows:

Lemma 4.1.1. *If the covering number $N(\epsilon, \mathcal{F}, d_P)$ is finite for every ϵ, $\mathcal{F}$
is PAC learnable. One can construct the so-called minimum risk algorithm
for which a number of $8/\epsilon^2 \cdot \ln (N(\epsilon/2, \mathcal{F}, d_P)/\delta)$ examples are sufficient for
a function class and a number of $32/\epsilon \cdot \ln (N(\epsilon/2, \mathcal{F}, d_P)/\delta)$ examples for a
concept class to PAC learn $\mathcal{F}$ with accuracy ϵ and confidence δ.*

*If $\mathcal{F}$ is a concept class, finiteness of the covering number for every $\epsilon > 0$ is
equivalent to PAC learnability. Any algorithm h which is PAC with accuracy
ϵ and confidence δ requires at least $\lg(M(2\epsilon, \mathcal{F}, d_P)(1 - \delta))$ examples.*

See [132] (Theorems 6.3, 6.4, and 6.5).

Considering real valued function classes, it is possible to specify the function that has to be learned by one output value uniquely, even if the function class is not countable. Due to this possibility examples of PAC learnable function classes with infinite covering number exist, *e.g.*, [132] (Example 6.11). But the necessity of a finite covering number holds even for functions with outputs in a finite but not necessarily binary set [132] (Theorem 6.7).

For both functions and concepts the property of consistent PUAC learnability can be characterized as follows:

Lemma 4.1.2. *The fact that $\mathcal{F}$ is consistently PUAC learnable is equivalent to the shrinking width property, that is, for any $\epsilon > 0$ and $\delta > 0$ a number $m_0 \in \mathbb{N}$ exists such that for every $m \geq m_0$*

$$P^m(\mathbf{x} \mid \exists f, g \, (\hat{d}_m(f, g, \mathbf{x}) = 0 \wedge d_P(f, g) > \epsilon)) \leq \delta.$$

See [132] (Theorem 6.2).

Finally, even for the UCED property one can find more appropriate characterizations. In a first step, the distances of two functions can be substituted by the mean of only one function, that is, the distance between the function and the constant function 0. The convergence of the empirical mean can also be correlated to covering numbers.

Lemma 4.1.3. *Assume that $\mathcal{F}$ has the property of uniform convergence of empirical means, or UCEM for short, that is, for any $\epsilon > 0$ and $\delta > 0$ a number $m_0 \in \mathbb{N}$ exists with*

$$P^m(\mathbf{x} \mid \sup_{f \in \mathcal{F}} |d_P(f, 0) - \hat{d}_m(f, 0, \mathbf{x})| > \epsilon) \leq \delta$$

for every $m \geq m_0$. Then it has the UCED property, too.

The UCEM property is equivalent to the condition that for any $\epsilon > 0$ and $\delta > 0$ some $m_0 \in \mathbb{N}$ exists such that for every $m \geq m_0$

$$\frac{\mathrm{E}_{P^m} \left(\lg(N(\epsilon, \mathcal{F}|\mathbf{x}, \hat{d}_m)) \right)}{m} \leq \delta,$$

where $\hat{d}_m$ is a short notation for the pseudometric measuring the empirical distance of two functions $\hat{d}_m(f, g, \mathbf{x}) = 1/m \sum_{i=1}^{m} |f(x_i) - g(x_i)|$ for functions f and g in $\mathcal{F}|\mathbf{x}$, and E_{P^m} denotes the expected value with respect to P^m. The following inequality holds:

$$P^m(\mathbf{x} \mid \sup_{f, g \in \mathcal{F}} |d_P(f, g) - \hat{d}_m(f, g, \mathbf{x})| > \epsilon)$$
$$\leq 2\mathrm{E}_{P^{2m}} (2 \, N(\epsilon/16, \mathcal{F}|\mathbf{x}, \hat{d}_{2m})^2) e^{-m\epsilon^2/32}$$

See [132] (Example 5.5, Corollary 5.6, Theorem 5.7).

Obviously, the UCEM property follows from the UCED property if the constant function 0 is contained in $\mathcal{F}$. These results establish properties that guarantee a positive answer to the questions 1-3 at the beginning of this chapter. But the bounds are still distribution-dependent and for the computation of these bounds it is necessary to estimate the covering number which depends on P.

4.1.2 Distribution-independent, Model-dependent Learning

In the *distribution-independent setting* nothing is known about the underlying probability measure P but *a priori* bounds on the number of examples necessary for valid generalization are to be established as well. Fortunately, one can limit the above distribution-dependent covering numbers by combinatorial bounds. These are independent of P such that the above inequalities hold even if the terms in the definition of PAC, PUAC, or UCED, respectively, are prefixed by a $\sup_P$, which means that the obtained inequalities hold regardless of the special probability measure P on X. For this purpose a quantity that measures the capacity of a concept or function class is introduced.

Definition 4.1.3. *Let $\mathcal{F}$ be a concept class. A set of points $\{x_1, \ldots, x_m\} \subset X$ is said to be* shattered *by $\mathcal{F}$ if for every mapping $d : \{x_1, \ldots, x_m\} \to \{0,1\}$ some $f \in \mathcal{F}$ exists with $f|\mathbf{x} = d$. The Vapnik Chervonenkis dimension $\mathcal{VC}(\mathcal{F})$ is the largest size of a set (maybe ∞) that is shattered by $\mathcal{F}$.*

Let $\mathcal{F}$ be a function class. A set of points $\{x_1, \ldots, x_m\} \subset X$ is said to be ϵ-fat shattered by $\mathcal{F}$ if real values $r_1, \ldots, r_m$ exist such that for every mapping $d : \{x_1, \ldots, x_m\} \to \{0,1\}$ some function $f \in \mathcal{F}$ exists with $d(x_i) = H(f(x_i) - r_i)$ and $|f(x_i) - r_i| \geq \epsilon$ for every i. For $\epsilon = 0$ we say 'shattered', too. The ϵ-fat shattering dimension $fat_\epsilon(\mathcal{F})$ is the largest size of a set that is ϵ-fat shattered by $\mathcal{F}$. For $\epsilon = 0$ this quantity is called the pseudo-dimension $\mathcal{PS}(\mathcal{F})$.

For $d = \mathcal{PS}(\mathcal{F})$ or $d = \mathcal{VC}(\mathcal{F})$, respectively, $d \geq 2$, and $\epsilon < 0.94$ it holds that

$$M(\epsilon, \mathcal{F}, d_P) \leq 2 \left(\frac{2e}{\epsilon} \ln \frac{2e}{\epsilon} \right)^d$$

for an arbitrary P [132] (Theorem 4.2). Consequently, finite VC- or pseudo-dimension, respectively, ensure learnability. Concrete bounds on the VC- or pseudo-dimension enable bounds on the convergence of a learning algorithm. Furthermore, for $d = \mathcal{VC}(\mathcal{F})$ and $\epsilon < 1/d$ a probability measure P exists such that $M(\epsilon, \mathcal{F}, d_P) \geq 2^d$; take the uniform distribution on d points that are shattered, for example. The argumentation can be improved such that a bound $M(\epsilon, \mathcal{F}, d_P) \geq e^{O(d)}$ arises for $\epsilon < 0.5$ [132] (Lemma 7.2). In particular, finiteness of the VC-dimension is both necessary and sufficient for distribution-independent PAC learnability of a concept class. In [13] it is shown that finiteness of the pseudo-dimension is necessary for the learnability of function classes with a finite number of outputs, too.

If $d = \mathcal{VC}(\mathcal{F})$ or $\mathcal{PS}(\mathcal{F})$ is finite, one can bound the covering number $N(\epsilon, \mathcal{F}|\mathbf{x}, \hat{d}_m)$ in Lemma 4.1.3 by $2((2e)/\epsilon \ln((2e)/\epsilon))^d$ ([132] Corollary 4.2). This leads to a number of

$$
O\left(\frac{d}{\epsilon} \ln \frac{1}{\epsilon} \cdot \ln\left(\ln \frac{1}{\epsilon}\right) + \frac{1}{\epsilon \delta}\right)
$$

examples which are sufficient such that every consistent algorithm is PUAC with accuracy ϵ and confidence δ [132] (Theorem 7.5). Furthermore, the UCED property is also guaranteed. In particular, the three terms PAC, PUAC, and UCED coincide in the distribution-independent setting for a concept class and can be characterized by the VC-dimension.

One can further bound the expected covering number by the inequality

$$
E_{P^m}(N(\epsilon, \mathcal{F}|\mathbf{x}, \hat{d}_m)) \le 2\left(\frac{4m}{\epsilon^2}\right)^{d \ln(2\,e\,m/(d\epsilon))}
$$

for function learning, where $d = fat_{\epsilon/4}(\mathcal{F})$ is the fat shattering dimension [2]. In fact, even the number $\sup_{\mathbf{x}} N(\epsilon, \mathcal{F}|\mathbf{x}, \hat{d}_m)$ is bounded by the above term. Consequently, the weaker condition of finite fat shattering dimension is sufficient for the learnability of functions. A number of $O(d/\epsilon^2 \cdot \ln(\ln(d/\epsilon)) + 1/\epsilon^2 \cdot \ln(1/\delta))$ examples is sufficient for the UCEM property where $d = fat_{\epsilon/24}(\mathcal{F})$ [2]. See [2] (Example 2.1) for a function class with infinite pseudo-dimension but finite fat shattering dimension for all ϵ. One can show that this condition is necessary for the UCEM property [2], and necessary for function learning under the presence of noise which fulfills some regularity conditions [7]. Because of this fact, the fat shattering dimension characterizes distribution-independent learnability of function classes under realistic conditions. Both here and in the concept case PUAC learnability and the UCED property come for free.

4.1.3 Model-free Learning

We have assumed that the function that has to be learned is itself contained in $\mathcal{F}$. This assumption is in general unrealistic since we know nothing about the underlying regularity. Therefore one can weaken the definition of learnability in the *model-free setting* as follows: Assume the unknown function f_0 is contained in a set $\mathcal{F}_0$, which may be different from $\mathcal{F}$. Then we can try to find a function in $\mathcal{F}$ approximating f_0 best. The minimum error achievable in $\mathcal{F}$ is of size

$$
J_P(f_0) = \inf_{f \in \mathcal{F}} d_P(f, f_0) \,.
$$

Given a sample $(x_i, f(x_i))_i$ the minimum empirical error achievable in $\mathcal{F}$ has the size

$$\hat{J}_m(f_0, \mathbf{x}) = \inf_{f \in \mathcal{F}} \sum_i |f_0(x_i) - f(x_i)|.$$

We define that $(\mathcal{F}_0, \mathcal{F})$ is PAC if an algorithm h exists such that for any $\epsilon > 0$ and $\delta > 0$ a number $m_0 \in \mathbb{N}$ exists such that for all $m \geq m_0$

$$\sup_{f_0 \in \mathcal{F}_0} P^m(\mathbf{x} \mid d_P(f_0, h_m(f_0, \mathbf{x})) - J_P(f_0) > \epsilon) \leq \delta.$$

PUAC learnability is characterized by the property that for any $\epsilon > 0$ and $\delta > 0$ a number $m_0 \in \mathbb{N}$ can be found such that for all $m \geq m_0$

$$P^m(\mathbf{x} \mid \sup_{f_0 \in \mathcal{F}_0} (d_P(f_0, h_m(f_0, \mathbf{x})) - J_P(f_0)) > \epsilon) \leq \delta.$$

The definition of the UCEM property need not be changed. As before, finiteness of the covering number of the approximating class $\mathcal{F}$ ensures PAC learnability [132] (Theorem 6.9). The UCEM property of $\mathcal{F}$ ensures in the distribution-independent case that every algorithm with empirical error converging to $\hat{J}_m(f_0, \mathbf{x})$ is PAC [132] (Theorems 3.2 and 5.11). The definition of UCEM is not different in the model-free case, hence this property can be correlated to the VC-, pseudo-, or fat shattering dimension leading to distribution-independent bounds for learnability in the model-free setting, too.

Note that this definition of model-free learning, which is a special case of the notation used in [132], fits only to a restricted setting. In practical applications one usually deals with noisy data, that is, probability distributions rather than simple functions have to be learned. In this situation the empirical error of a probability P_f on $X \times [0, 1]$ compared to a function g may be defined as $d_P(P_f, g) = \int_{X \times [0,1]} |y - f(x)| dP_f(x, y)$. Furthermore, this measure may be not appropriate since some deviations of the output, from the function that has to be learned, may be worse than others. For example, a substitution of the Euclidian distance by the quadratic error punishes large deviations more than small ones. All these modifications fit into the framework of agnostic learning, as established in the work of Haussler [51], who has generalized the approach of Vapnik and Chervonenkis [12, 131] about concept classes to real valued functions.

Note that the explicit bounds on the number of examples can be improved for concept learning [132]. Furthermore, other useful modifications of the entire learning scenario exist which we will not consider in the following, see [5, 52, 74, 75, 94, 95, 133].

As a consequence of these results, it is appropriate to first consider the VC-, pseudo-, or fat shattering dimension when examining the learnability of a function class. If this dimension is finite, learnability can be guaranteed and explicit bounds on the number of examples can be established. If this dimension is infinite or very large so that prohibitive bounds on the number of examples result, one can either restrict the function class to a class with smaller VC-dimension or try to estimate the covering number for the special distribution – maybe these methods lead to better bounds.

4.1.4 Dealing with Infinite Capacity

In fact, when considering the entire story one often deals with a class of infinite VC-dimension. This is the natural setting if a function with an unknown structure is to be learned. To fit a model to such a function it is necessary to start with a class which has some kind of universal approximation capability – otherwise the approximation process would fail in at least some situations. We have proved the universal approximation capability of folding networks in the previous chapter and it has also been established for standard feed-forward networks [59]. Since any function on vectors, lists, or trees, respectively, can be represented by the class of feed-forward, recurrent, or folding networks these classes obviously have infinite capacity.

One usual way to overcome the problem of infinite VC-dimension in a concrete learning scenario is due to the fact that the function class is structured in a natural way by the number of parameters that can be fitted. In neural network learning first an architecture is chosen and then the network parameters are trained. Generally speaking, when approximating an unknown function one can first estimate the complexity and rough structure of the function so that the maximum number of parameters can be limited. Afterwards, we can learn with this class with a finite number of parameters and finite capacity and obtain guarantees for the generalization ability, as already described in this chapter. In fact, this procedure first minimizes the so-called structural risk by restricting the capacity – fixing the neural architecture – and it afterwards minimizes the empirical risk by fitting the parameters to the data – back-propagation, for example. The error of the approximation is then bounded by the sum of the structural risk, that is, the deviation of the real error and the training error, and the empirical risk, that is, the training error [130]. The empirical risk can be measured directly, we have to evaluate the output function of the training algorithm at all training examples. The structural risk can be estimated by the bounds which depend on the VC- or pseudo-dimension or related terms we have considered in this chapter. Note that these two minimization procedures, the empirical risk minimization and the structural risk minimization, are contradictory tasks because the structural risk is smaller for a parameterized subset with small capacity and a large amount of training data, whereas the empirical risk becomes worse in this situation. Therefore an appropriate balance between these two tasks has to be found.

Instead of the above-mentioned method a component which further reduces the structural risk is often added to the empirical risk minimization in neural network learning: A weight decay or some other penalty term is added to the empirical error which has to be minimized, for example, and corresponds to a regularization of the function [18]. Methods even exist where the structural risk minimization is performed after the empirical risk minimization, as in Vapnik's support vector method [23, 130]. To take this method into account some argumentation has to be added to the theoretical analysis, for example, as we will now discuss, the luckiness framework.

Now, when dealing with recurrent networks the above method – estimating the number of parameters and fitting the data afterwards – is not applicable because the hierarchy described by the number of parameters collapses when considering the corresponding complexity of the classes, as we will see in the next section. It will be shown that even neural architectures with a finite and actually very small number of parameters have unlimited capacity. Consequently, one has to add a component to the theoretical analysis to ensure learnability in this case, too. The hierarchy defined by the number of parameters has to be further refined in some way.

Two approaches in the latter case may be useful: Assume the VC-dimension of a concept class $\mathcal{F}$ is infinite, but the input space X can be written as a union of subspaces $\bigcup_{i=1}^{\infty} X_i$ such that $X_i \subset X_{i+1}$ and $\mathcal{VC}(\mathcal{F}|X_i) = i$ for all i. Then every consistent algorithm is PAC and requires only a polynomial number of examples, provided that $1 - P(X_i) = O(i^{-\beta})$ for some $\beta > 0$ [3]. The division of the input space will turn out to be rather natural when dealing with recursive data. One can consider the sets of trees where the maximum height is restricted, for example. But at least some prior information about the probability measure – the probability of high trees – is necessary.

Another approach deals with the special output of a learning algorithm on the real data and guarantees a certain accuracy that is dependent on the concept class as well as on the luckiness of the function which is the concrete output of the algorithm. For this purpose a *luckiness function* $L : X^m \times \mathcal{F} \to \mathbb{R}^+$ is given. This luckiness function measures the luckiness of the function approximating the data which is the output of the learning algorithm. The luckiness may measure the margin or number of support vectors in the case of classification with one hyperplane, for example. The luckiness tells us whether the output function lies in a function class with small capacity, that is, small VC- or pseudo-dimension. For technical reasons the luckiness of a function on a double sample has to be estimated on the first half of the sample in the following sense: Functions η and $\phi : \mathbb{N} \times \mathbb{R}^+ \times \mathbb{R}^+ \to \mathbb{R}^+$ exist such that for any $\delta > 0$

$$\sup_{f \in \mathcal{F}} P^{2m}(\mathbf{xy} \mid \exists g \in \mathcal{F} \, (\hat{d}_m(f, g, \mathbf{x}) = 0 \wedge \forall \mathbf{x'y'} \subset_\eta \mathbf{xy}$$
$$|\{h|\mathbf{x'y'} \mid h \in \mathcal{F} \wedge L(\mathbf{x'y'}, h) \geq L(\mathbf{x'y'}, g)\}| > \phi(m, L(\mathbf{x}, g), \delta))) \leq \delta$$

where $\mathbf{x'y'} \subset_\eta \mathbf{xy}$ refers to any vector which results if a fraction of $\eta = \eta(m, L(\mathbf{x}, g), \delta)$ coefficients are deleted in the part $\mathbf{x}$ and in the part $\mathbf{y}$ of the vector $\mathbf{xy}$. The condition requires some kind of smoothness of the luckiness function, *i.e.*, if we know how lucky we are in a certain training situation then we can estimate the number of situations which would be at least as lucky even if more training data would be available. Now if an algorithm outputs on m i.i.d. examples a concept $h = h_m(f, \mathbf{x})$, which is consistent with the examples such that $\phi(m, L(\mathbf{x}, h), \delta) \leq 2^{i+1}$, then with probability of at least $1 - \delta$ the following inequality holds for any probability P:

$$d_P(f,h) \leq \frac{2}{m}\left(i+1+\lg\frac{4}{p_i\delta}\right) + 4\eta\left(m, L(\mathbf{x},h), \frac{p_i\delta}{4}\right)\lg(4m),$$

where p_i are positive numbers satisfying $\sum_{i=1}^{2m} p_i = 1$ [113]. Note that this approach leads to posterior bounds, although the p_i represent in some way the confidence of getting an output with a certain luckiness. This approach will be used to estimate the probability of a certain height of the input trees.

4.1.5 VC-Dimension of Neural Networks

We want to apply these theoretical results to neural network learning. For this purpose, bounds on the VC-dimension of neural architectures are needed. Denote by W the number of weights in an architecture including biases, by N the number of neurons. For feed-forward architectures with activation function σ the following upper, lower, or sharp bounds, respectively, have been established for the VC- or pseudo-dimension d:

$$d = \begin{cases} \Theta(W \ln W) & \text{if } \sigma = \text{H,} \\ O(W^2 N^2) & \text{if } \sigma = \text{sgd,} \\ \Omega(WN) & \text{if } \sigma = \text{sgd,} \\ O(WN \ln q + Wh \ln d) & \text{if } \sigma \text{ is piecewise polynomial,} \\ & q \text{ is the maximum number of pieces,} \\ & d \text{ is the maximum degree,} \\ & \text{and } h \text{ is the depth of the architecture.} \end{cases}$$

See [12, 64, 69, 82, 83, 118, 132].

In the last three cases weight sharing is allowed and W denotes only the number of different adjustable parameters. Note that in the last estimation the term $\ln d^h$ is an upper bound for the maximum degree of a polynomial with respect to the weights, which occurs in a formula corresponding to the network computation. If the activation function σ is linear this degree is h, which leads to an improvement of this factor to $\ln h$.

Consequently, in interesting cases good upper bounds exist and learnability is established for feed-forward architectures with a standard activation function. Furthermore, the approach [86] proves the finiteness of the VC-dimension and consequently the learnability if the activation function is an arbitrary algebraic function. On the contrary, an activation function $\sigma = \cos$ can lead to an infinite VC-dimension. One can even construct activation functions which look very similar to the standard sigmoidal function such that very small architectures have an infinite VC-dimension due to a hidden oscillation of the activation function [119]. But in both cases a restriction of the weights limits the capacity, too, because in this case the oscillation is limited.

The above bounds on the VC-dimension of neural architectures can be improved if networks with a finite input set or a fixed number of layers are considered [8, 9]. Furthermore, in [10] the fat shattering dimension of feed-forward architectures is examined. It turns out that good upper bounds can be derived for networks with small weights and depths.

Some approaches deal with the VC-dimension of recurrent architectures where the corresponding feed-forward architecture has depth 1 and derive the bounds

$$d = \begin{cases} O(W^2 t) & \text{if } \sigma \text{ is piecewise polynomial,} \\ O(Wt) & \text{if } \sigma \text{ is a polynomial,} \\ O(W \ln t) & \text{if } \sigma \text{ is linear,} \\ O(WN + W \ln(Wt)) & \text{if } \sigma = \text{H,} \\ O(W^2 N^2 t^2) & \text{if } \sigma = \text{sgd,} \\ \Omega(W \ln(t/W)) & \text{if } \sigma = \text{H or } \sigma = \text{id,} \\ \Omega(Wt) & \text{if } \sigma = \text{sgd or a nonlinear polynomial,} \end{cases}$$

where t is the maximum input length of an input sequence [27, 70]. Since the lower bounds depend on t and become infinite for arbitrary t distribution-independent PAC learnability of recurrent architectures cannot be guaranteed in general. The hierarchy which is defined by the number of parameters collapses when considering the corresponding capacities.

4.2 PAC Learnability

In this section we present some general results concerning PAC learnability which deal with the learning of recursive data and are therefore of interest when training folding architectures, or which are interesting for the possibility of learning under certain conditions in general.

4.2.1 Distribution-dependent Learning

First of all, we consider the distribution-dependent setting. Here the term PUAC has been introduced by Vidyasagar [132]. The stronger condition of uniform convergence is fulfilled automatically in the distribution-independent case if only PAC learnability is required. In [132] (Problem 12.4) the question is posed whether these terms coincide in the distribution-dependent case, too. Indeed, this is not the case, as can be seen by example:

Example 4.2.1. A concept class exists that is PAC, but not PUAC learnable in the distribution-dependent setting.

Proof. Consider $X = [0, 1]$, $\mathcal{P} =$ uniform distribution and $\mathcal{F} = \{f : [0, 1] \to \{0, 1\} \mid f \text{ is constant almost surely}\}$. Since $\mathcal{F}$ has a finite covering for every $\epsilon > 0$, it is PAC learnable. Assume an algorithm h exists that is PUAC. Assume that for a sample $\mathbf{x}$ and function f the function $h_m(f, \mathbf{x})$ is almost surely 0. Then for the function g, which equals f on $\mathbf{x}$ and is almost surely 1, the distance $d_P(g, h_m(g, \mathbf{x}))$ is 1 because $h_m(f, \mathbf{x}) = h_m(g, \mathbf{x})$. An analogous situation holds if h produces a function that is almost surely 1. Therefore $P(\mathbf{x} \mid \sup_{f \in \mathcal{F}} d_P(f, h_m(f, \mathbf{x})) > \epsilon) = 1$. □

What is the advantage of the requirement of PUAC learnability in distribution-dependent learning? One nice property is that with PUAC learnability we get consistent PUAC learnability for free.

Theorem 4.2.1. *$\mathcal{F}$ is PUAC learnable if and only if $\mathcal{F}$ is consistently PUAC learnable.*

Proof. Assume that not every consistent algorithm is PUAC and therefore the shrinking width property is violated. For an arbitrary learning algorithm h, a sample $\mathbf{x}$ and functions f and g with $\hat{d}_m(f, g, \mathbf{x}) = 0$, it is valid that $d_P(f, g) \leq d_P(f, h_m(f, \mathbf{x})) + d_P(g, h_m(g, \mathbf{x}))$. If $d_P(f, g) > \epsilon$, at least one of the functions f and g has a distance of at least $\epsilon/2$ from the function produced by the algorithm. Therefore $\{\mathbf{x} \mid \exists f, g \, (\hat{d}_m(f, g, \mathbf{x}) = 0 \wedge d_P(f, g) > \epsilon)\} \subset \{\mathbf{x} \mid \sup_{f \in \mathcal{F}} d_P(f, h_m(f, \mathbf{x})) > \epsilon/2\}$; the probability of the latter set is at least as large as the probability of the first one. That is, both probabilities do not tend to 0 for increasing m because the shrinking width property is violated. □

As a consequence any consistent algorithm is appropriate if the PUAC property holds for the function class. Furthermore, a characterization of PUAC learnability which does not refer to the notion of a concrete learning algorithm is given by the shrinking width property.

On the contrary, PAC learnability and consistent PAC learnability are different concepts, as can be seen by the following example:

Example 4.2.2. A PAC learnable concept class exists for which not every consistent algorithm is PAC.

Proof. Consider the following scenario: $X = [0, 1]$, $P =$ uniform distribution, $\mathcal{F} = \{f : X \to \{0, 1\} \mid f(x) = 0 \text{ almost surely or } f(x) = 1 \text{ for all } x\}$. Consider the algorithm which produces the function 1 until a value x with image 0 is an element of the sample. The algorithm is consistent and PAC because for any function except 1 the set with image 1 has the measure 0. However, the consistent algorithm which produces the function with image 1 on only the corresponding elements of the sample and 0 on any other point is not PAC because the constant function 1 cannot be learned. □

That means, PAC learnability ensures the existence of at least one good learning algorithm but some consistent algorithms may fail to learn a function correctly. Of course, the difference between PAC and PUAC follows from this abstract property, too.

It would be nice to obtain a characterization of consistent PAC learnability which does not refer to the notion of a learning algorithm as well. The following condition requires the possibility that any function can be characterized by a finite set of points:

Theorem 4.2.2. *A function class is consistently PAC learnable if and only if it is* finitely characterizable, *that is, for all $\epsilon > 0$ and $\delta > 0$ a number $m_0 \in \mathbb{N}$ exists such that for all $m \geq m_0$*

$$\sup_{f \in \mathcal{F}} P^m(\mathbf{x} \mid \exists g \, (\hat{d}_m(f, g, \mathbf{x}) = 0 \wedge d_P(f, g) > \epsilon)) \leq \delta.$$

Proof. Assume that $\mathcal{F}$ is finitely characterizable and the algorithm h is consistent. The following inequality bounds the error of the algorithm h:

$$\begin{aligned}
&\sup_f P^m(\mathbf{x} \mid d_P(f, h_m(f, \mathbf{x}), \mathbf{x}) > \epsilon) \\
\leq\ &\sup_f P^m(\mathbf{x} \mid \hat{d}_m(f, h_m(f, \mathbf{x}), \mathbf{x}) \neq 0) \\
&+ \sup_f P^m(\mathbf{x} \mid d_P(f, h_m(f, \mathbf{x}), \mathbf{x}) > \epsilon \text{ and } \hat{d}_m(f, h_m(f, \mathbf{x}), \mathbf{x}) = 0) \\
\leq\ &\sup_f P^m(\mathbf{x} \mid \hat{d}_m(f, h_m(f, \mathbf{x}), \mathbf{x}) \neq 0) \\
&+ \sup_f P^m(\mathbf{x} \mid \exists g \, (d_P(f, g, \mathbf{x}) > \epsilon \text{ and } \hat{d}_m(f, g, \mathbf{x}) = 0)).
\end{aligned}$$

If conversely the condition of finite characterizability is violated, some $\epsilon > 0$, $\delta > 0$, numbers $n_1, n_2, \ldots \to \infty$, and functions $f_1, f_2, \ldots$ exist such that

$$P^{n_i}(\mathbf{x} \mid \exists g \, (\hat{d}_{n_i}(f_i, g, \mathbf{x}) = 0 \wedge d_P(f_i, g) > \epsilon)) > \delta.$$

Choose for every $\mathbf{x}$ from the support of the above set one $g_\mathbf{x}^i$ such that the properties $\hat{d}_{n_i}(f_i, g_\mathbf{x}^i, \mathbf{x}) = 0$ and $d_P(f_i, g_\mathbf{x}^i) > \epsilon$ hold. The partial mapping $(\mathbf{x}, f_i(\mathbf{x})) \mapsto g_\mathbf{x}^i$ can be completed to a consistent learning algorithm h which is not PAC. $\qquad\square$

Note that the above argumentation shows additionally that the condition of finite characterizability implies that any asymptotically consistent algorithm is PAC, too. 'Asymptotically consistent' means that the probability of points where the empirical error does not vanish tends to zero. In particular, finite characterizability is a desirable property of a function class which is a weaker condition than PUAC learnability:

Example 4.2.3. A concept class exists which is consistently PAC learnable but not PUAC learnable.

Proof. Consider

$$\begin{aligned}
\mathcal{F} = \{\ & f : [0, 1] \to \{0, 1\} \mid \\
& (f \text{ is } 1 \text{ almost surely on } [0, 0.5[\text{ and } f \text{ is } 1 \text{ on } [0.5, 1]) \\
& \text{or } (f \text{ is } 0 \text{ on } [0, 0.5[\text{ and } f \text{ is } 0 \text{ almost surely on } [0.5, 1]) \}
\end{aligned}$$

and the uniform distribution on $[0, 1]$. $\sup_f P^m(\mathbf{x} \mid \exists g \, (\hat{d}_m(f, g, \mathbf{x}) = 0 \wedge d_P(f, g) > \epsilon)) \to 0$ is valid because for a function which is 1 almost surely, for example, the set of points in $[0, 0.5[$ where f is 1 has measure 0.5. These points characterize f. On the contrary, $P^m(\mathbf{x} \mid \exists f, g \, (\hat{d}_m(f, g, \mathbf{x}) =$

$0 \wedge d_P(f,g) > \epsilon)) = 1$ because we can find for any $\mathbf{x}$ functions f and g which are 1 or 0 almost surely, respectively, but $f|(\{x_1, \ldots, x_m\} \cap [0, 0.5[) = g|(\{x_1, \ldots, x_m\} \cap [0, 0.5[) = 0$ and analogous for $[0.5, 1]$. Consequently, $\mathcal{F}$ is consistently PAC learnable but not PUAC. $\square$

4.2.2 Scale Sensitive Terms

The above characterizations are not fully satisfactory since a concrete learning algorithm often only produces a solution which merely minimizes the empirical error rather than bringing the error to 0. This can be due, for example, to the complexity of the empirical error minimization. In the case of model-free learning, which is considered later, it is possible that a minimum simply does not exist. Therefore the above terms are weakened in the following definition.

Definition 4.2.1. *An algorithm h is* asymptotically ϵ-consistent *if for any $\delta > 0$ a number $m_0 \in \mathbb{N}$ exists such that for all $m \geq m_0$*

$$\sup_{f \in \mathcal{F}} P^m(\mathbf{x} \mid \hat{d}_m(f, h_m(f, \mathbf{x}), \mathbf{x}) > \epsilon) \leq \delta.$$

h is asymptotically uniformly ϵ-consistent *if for any $\delta > 0$ a number $m_0 \in \mathbb{N}$ exists such that for all $m \geq m_0$*

$$P^m(\mathbf{x} \mid \sup_{f \in \mathcal{F}} \hat{d}_m(f, h_m(f, \mathbf{x}), \mathbf{x}) > \epsilon) \leq \delta.$$

A function class $\mathcal{F}$ is finitely ϵ_1-ϵ_2-characterizable *if for any $\delta > 0$ some $m_0 \in \mathbb{N}$ can be found such that for all $m \geq m_0$*

$$\sup_{f \in \mathcal{F}} P^m(\mathbf{x} \mid \exists g \, (\hat{d}_m(f, g, \mathbf{x}) \leq \epsilon_1 \wedge d_P(f, g) > \epsilon_2)) \leq \delta.$$

$\mathcal{F}$ fulfills the ϵ_1-ϵ_2-shrinking width property if for any $\delta > 0$ some $m_0 \in \mathbb{N}$ exists such that for all $m \geq m_0$

$$P^m(\mathbf{x} \mid \exists f, g \, (\hat{d}_m(f, g, \mathbf{x}) \leq \epsilon_1 \wedge d_P(f, g) > \epsilon_2)) \leq \delta.$$

$\mathcal{F}$ is ϵ_1-consistently PAC learnable with accuracy ϵ_2 if any asymptotically ϵ_1-consistent algorithm is PAC with accuracy ϵ_2.
$\mathcal{F}$ is ϵ_1-consistently PUAC learnable with accuracy ϵ_2 if any asymptotically uniformly ϵ_1-consistent algorithm is PUAC with accuracy ϵ_2.

In the case $\epsilon = 0$ and $\epsilon_1 = 0$ the previous definitions result with the difference that here, the algorithms are only required to have small or zero error in the limit. Now we consider the question as to whether an algorithm which minimizes the empirical error with a certain degree is PAC with a certain accuracy. Analogous to the case of a consistent algorithm the following theorem holds.

Theorem 4.2.3. $\mathcal{F}$ *is* ϵ_1-*consistently PAC learnable with accuracy* ϵ_2 *if and only if* $\mathcal{F}$ *is finitely* ϵ_1-ϵ_2-*characterizable.*
$\mathcal{F}$ *is* ϵ_1-*consistently PUAC learnable with accuracy* ϵ_2 *if and only if* $\mathcal{F}$ *fulfills the* ϵ_1-ϵ_2-*shrinking width property.*

Proof. Assume that $\mathcal{F}$ is finitely ϵ_1-ϵ_2-characterizable and the algorithm h is ϵ_1-consistent. The following inequality bounds the error of the algorithm h:

$$\sup_f P^m(\mathbf{x} \mid d_P(f, h_m(f, \mathbf{x}), \mathbf{x}) > \epsilon_2)$$
$$\leq \quad \sup_f P^m(\mathbf{x} \mid \hat{d}_m(f, h_m(f, \mathbf{x}), \mathbf{x}) > \epsilon_1)$$
$$\quad + \sup_f P^m(\mathbf{x} \mid d_P(f, h_m(f, \mathbf{x}), \mathbf{x}) > \epsilon_2 \text{ and } \hat{d}_m(f, h_m(f, \mathbf{x}), \mathbf{x}) \leq \epsilon_1)$$
$$\leq \quad \sup_f P^m(\mathbf{x} \mid \hat{d}_m(f, h_m(f, \mathbf{x}), \mathbf{x}) > \epsilon_1)$$
$$\quad + \sup_f P^m(\mathbf{x} \mid \exists g\,(d_P(f, g, \mathbf{x}) > \epsilon_2 \text{ and } \hat{d}_m(f, g, \mathbf{x}) \leq \epsilon_1)).$$

If conversely, the condition of finite ϵ_1-ϵ_2-characterizability is violated, some $\delta > 0$, numbers $n_1, n_2, \ldots \to \infty$, and functions $f_1, f_2, \ldots$ exist such that

$$P^{n_i}(\mathbf{x} \mid \exists g\,(\hat{d}_{n_i}(f_i, g, \mathbf{x}) \leq \epsilon_1 \wedge d_P(f_i, g) > \epsilon_2) > \delta.$$

Choose for every $\mathbf{x}$ from the support of the above set one $g_{\mathbf{x}}^i$ such that the properties $\hat{d}_{n_i}(f_i, g_{\mathbf{x}}^i, \mathbf{x}) \leq \epsilon_1$ and $d_P(f_i, g_{\mathbf{x}}^i) > \epsilon_2$ hold. The partial mapping $(\mathbf{x}, f_i(\mathbf{x})) \mapsto g_{\mathbf{x}}^i$ can be completed to a ϵ_1-consistent learning algorithm h which is not PAC with accuracy ϵ_2. This shows the first half of the theorem.

In the uniform case, assume that $\mathcal{F}$ possesses the ϵ_1-ϵ_2-shrinking width property and the algorithm h is ϵ_1-consistent. The following inequality bounds the error of the algorithm h:

$$P^m(\mathbf{x} \mid \exists f\, d_P(f, h_m(f, \mathbf{x}), \mathbf{x}) > \epsilon_2)$$
$$\leq \quad P^m(\mathbf{x} \mid \exists f\, \hat{d}_m(f, h_m(f, \mathbf{x}), \mathbf{x}) > \epsilon_1)$$
$$\quad + P^m(\mathbf{x} \mid \exists f\,(d_P(f, h_m(f, \mathbf{x}), \mathbf{x}) > \epsilon_2 \text{ and } \hat{d}_m(f, h_m(f, \mathbf{x}), \mathbf{x}) \leq \epsilon_1))$$
$$\leq \quad P^m(\mathbf{x} \mid \exists f\, \hat{d}_m(f, h_m(f, \mathbf{x}), \mathbf{x}) > \epsilon_1)$$
$$\quad + P^m(\mathbf{x} \mid \exists f, g\,(d_P(f, g, \mathbf{x}) > \epsilon_2 \text{ and } \hat{d}_m(f, g, \mathbf{x}) \leq \epsilon_1)).$$

If conversely, the ϵ_1-ϵ_2-shrinking width property is violated, some $\delta > 0$ and numbers $n_1, n_2, \ldots \to \infty$ exist such that

$$P^{n_i}(\mathbf{x} \mid \exists f, g\,(\hat{d}_{n_i}(f, g, \mathbf{x}) \leq \epsilon_1 \wedge d_P(f, g) > \epsilon_2) > \delta.$$

Choose for every $\mathbf{x}$ from the support of the above set $f_{\mathbf{x}}^i$ and $g_{\mathbf{x}}^i$ such that the properties $\hat{d}_{n_i}(f_{\mathbf{x}}^i, g_{\mathbf{x}}^i, \mathbf{x}) \leq \epsilon_1$ and $d_P(f_{\mathbf{x}}^i, g_{\mathbf{x}}^i) > \epsilon_2$ hold. The partial mapping $(\mathbf{x}, f_{\mathbf{x}}^i(\mathbf{x})) \mapsto g_{\mathbf{x}}^i$ can be completed to a ϵ_1-consistent learning algorithm h which is not PUAC with accuracy ϵ_2. $\quad\square$

Note that the above argumentation is very similar to the proof of Theorem 4.2.2. In fact, we could substitute the property $d_P(f, g) > \epsilon_2$ occurring in the definition of PAC or PUAC, respectively, by some abstract property E_1 and the property $\hat{d}_m(f, g, \mathbf{x}) \leq \epsilon_1$ which occurs in the definition of consistency by some abstract property E_2. Assume we want to guarantee that for any algorithm which fulfills $E_2(f, h_m(f, \mathbf{x}))$ with high probability automatically $E_1(f, h_m(f, \mathbf{x}))$ does not hold with high probability. The above argumentation shows that this implication is true if and only if the probability that E_1 and E_2 hold in common is small. Note that the latter characterization is independent of the notion of a learning algorithm. Depending on whether the probability is uniform or not the above characterizations: 'finitely characterizable' and 'shrinking width' result in our case. The only thing we have to take care of is that any partial learning algorithm which fulfills E_2 can be completed to a learning algorithm which still fulfills E_2.

The PUAC property no longer guarantees that any asymptotically ϵ_1-consistent algorithm is PUAC. An additional condition is needed.

Theorem 4.2.4. *An algorithm is called ϵ_1-ϵ_2-stable if for all $\delta > 0$ a number $m_0 \in \mathbb{N}$ exists such that for all $m \geq m_0$*

$$P^m(\mathbf{x} \mid \exists f, g\, (\hat{d}_m(f, g, \mathbf{x}) \leq \epsilon_1 \wedge d_P(h_m(f, \mathbf{x}), h_m(g, \mathbf{x})) > \epsilon_2)) \leq \delta.$$

A function class $\mathcal{F}$ is ϵ_1-consistently PUAC learnable with accuracy $3\epsilon_2$ if some PUAC algorithm with accuracy ϵ_2 exists which is ϵ_1-ϵ_2-stable.

Conversely, if $\mathcal{F}$ is ϵ_1-consistently PUAC learnable with accuracy ϵ_2 then any PUAC algorithm is ϵ_1-$3\epsilon_2$-stable.

Proof. For any learning algorithm h the following holds:

$$\begin{aligned}
&\{\mathbf{x} \mid \exists f, g\, (\hat{d}_m(f, g, \mathbf{x}) \leq \epsilon_1 \wedge d_P(f, g) > 3\epsilon_2)\} \\
\subset\ &\{\mathbf{x} \mid \exists f, g\, (\hat{d}_m(f, g, \mathbf{x}) \leq \epsilon_1 \wedge d_P(h_m(f, \mathbf{x}), h_m(g, \mathbf{x})) > \epsilon_2)\} \\
\cup\ &\{\mathbf{x} \mid \sup_f d_P(f, h_m(f, \mathbf{x})) > \epsilon_2\}.
\end{aligned}$$

For a PUAC and stable algorithm h the probability of the latter two sets tends to 0.
Conversely,

$$\begin{aligned}
&\{\mathbf{x} \mid \exists f, g\, (\hat{d}_m(f, g, \mathbf{x}) \leq \epsilon_1 \wedge d_P(h_m(f, \mathbf{x}), h_m(g, \mathbf{x})) > 3\epsilon_2)\} \\
\subset\ &\{\mathbf{x} \mid \exists f, g\, (\hat{d}_m(f, g, \mathbf{x}) \leq \epsilon_1 \wedge d_P(f, g) > \epsilon_2)\} \\
\cup\ &\{\mathbf{x} \mid \sup_f d_P(f, h_m(f, \mathbf{x})) > \epsilon_2\},
\end{aligned}$$

therefore the second statement also follows. $\qquad\qquad\square$

The stability criterion requires that small deviations in the input of an algorithm only lead to small deviations in the output function of this algorithm. The stability property is automatically fulfilled for any PUAC algorithm of a function class which has the UCED property because this property guarantees that a small empirical error is representative for the real error.

But concept classes exist which are ϵ-consistently PUAC learnable for every accuracy but do not have the UCED property. Consider, for example, the class

$$\mathcal{F} = \{f : [0,1] \to \{0,1\} \mid f \text{ is constant } 0 \text{ almost surely}\}$$

and the uniform distribution on $[0,1]$. The empirical error $\hat{d}_m(f,0,\mathbf{x})$ may be much larger than the real error $d_P(f,0)$, which is always 0.

Let us examine the relation between the other terms that have been introduced, too. The following diagram results where ϵ_1, ϵ_2, and ϵ_3 are assumed to be positive:

$\mathcal{F}$ is PAC learnable $\qquad\qquad$ $\mathcal{F}$ is PUAC learnable

$\Uparrow$ $\qquad\qquad\qquad\qquad\qquad$ $\Updownarrow$

$\mathcal{F}$ is consistently $\qquad\qquad$ $\mathcal{F}$ is consistently
PAC learnable $\quad\Leftarrow\quad$ PUAC learnable

$\Uparrow_{(1)}$ $\qquad\qquad\qquad\qquad\qquad$ $\Uparrow_{(2)}$

$\forall\epsilon_2\exists\epsilon_1$ such that $\mathcal{F}$ is ϵ_1- $\qquad$ $\forall\epsilon_2\exists\epsilon_1$ such that $\mathcal{F}$ is
consistently PAC learnable $\quad\Leftarrow_{(3)}\quad$ ϵ_1-consistently PUAC
learnable with accuracy ϵ_2 $\qquad$ with accuracy ϵ_2

$\Updownarrow$

$\forall\epsilon_2\forall\epsilon_3\exists\epsilon_1$ s.t. some ϵ_1-ϵ_3 stable
learning algorithm PUAC
learns $\mathcal{F}$ with accuracy ϵ_2

$\Uparrow$

$\mathcal{F}$ possesses the UCED property

All inclusions are strict. Furthermore, PUAC and ϵ-consistent PAC learnability are incomparable terms (which we refer to as (4)).

The strictness in (1)-(4) still has to be shown. Note that the counterexamples for the other inequalities have only used concept classes. Additionally, the concept class in Example 4.2.3 is even ϵ-consistently PAC learnable for every $0 < \epsilon < 0.5$ with any accuracy $0 < \epsilon_1 < 1$, which implies inequality (3) and one direction of (4). The remaining inequalities follow from the next example.

Example 4.2.4. A function class exists which is PUAC but not ϵ-consistently PAC learnable for any $\epsilon > 0$ and any accuracy $0 < \epsilon_1 < 1$.

Proof. Consider the function class $\mathcal{F} = \bigcup_{i=1}^{\infty} \mathcal{F}_i \cup \mathcal{F}_i'$, where

$$\mathcal{F}_i = \{f : [0,1] \to [0,1] \mid f(x) \in \{0, (1+e^{-i})^{-1}\}, f \text{ is } 0 \text{ almost surely}\}$$
$$\mathcal{F}_i' = \{f : [0,1] \to [0,1] \mid f(x) \in \{1, 1-(1+e^{-i})^{-1}\}, f \text{ is } 1 \text{ almost surely}\},$$

and consider the uniform distribution on $[0,1]$. These functions are PUAC learnable since the function values uniquely determine whether the function is 0 or 1 almost surely. On the contrary, for any function f, sample $\mathbf{x}$, and $\epsilon > 0$ a function exists with distance 1 and empirical distance smaller than ϵ from f. $\qquad\square$

Unfortunately, this example uses a function class with real outputs and the possibility of encoding a function uniquely into a single output value. Of course, this argument cannot be transferred to concepts. Considering concept classes the following example can be constructed:

Example 4.2.5. For any positive ϵ a concept class exists which is PUAC but not ϵ-consistently PAC learnable for any accuracy $\epsilon_1 < 1$.

Proof. Consider $\mathcal{F} = \{f : [0,1] \to \{0,1\} \mid f$ is 1 or $(f|[0, 1 - \epsilon/2[$ is 0 almost surely and $f|[1 - \epsilon/2, 1]$ is 0)$\}$ and the uniform distribution. Since for large samples nearly a fraction $\epsilon/2$ of the example points is contained in $[1 - \epsilon/2, 1]$, and therefore determine the function with high probability, this concept class is PUAC. On the contrary, we can find a function with empirical distance at most ϵ and real distance 1 from the constant function 1 for any $\mathbf{x}$ where only a fraction ϵ of the examples is contained in $[1 - \epsilon/2, 1]$. $\qquad\square$

This example does not answer the question as to whether a positive ϵ exists for any concept class which is PUAC learnable such that the class is ϵ-consistently PAC learnable, too. The example only shows that such an ϵ cannot be identical for all concept classes.

4.2.3 Noisy Data

The possibility that for function classes the entire function can be encoded in one real output value causes some difficulties in characterizing PAC learnability in general and prohibits an exact characterization of PAC learnable function classes. Of course, in practical applications the real numbers are only presented with a bounded accuracy, due to the bounded computational capacity. Furthermore, the presence of noise is an inherent property of a learning scenario which prohibits the exact encoding of functions. It has been shown that under the presence of noise, the learnability of a function class in the distribution-independent case reduces to the learnability of a finite valued class [7]. The argumentation can immediately be transferred to the distribution-dependent setting as follows: Let us first introduce some noise into the learning scenario.

Definition 4.2.2. *A randomized learning algorithm for a function class $\mathcal{F}$ is a mapping $h : \bigcup_{m=1}^{\infty} (X \times Y \times Z)^m \to \mathcal{F}$ together with a probability distribution P_Z on the measurable space Z. A randomized algorithm is PAC with accuracy ϵ and confidence δ on m examples if*

$$\sup_{f \in \mathcal{F}} P^m \times P_Z^m((\mathbf{x}, \mathbf{z}) \mid d_P(h_m(f, \mathbf{x}, \mathbf{z}), f) > \epsilon) \leq \delta.$$

The definitions of PUAC, consistent, ... transfer to a randomized algorithm in a similar way.

The purpose is that, together with the examples, a randomized algorithm can use random patterns that are taken according to P_Z. It may, for example, use a tossed coin if two or more functions of the class fit well to the data. At least for finite valued function classes this notation does not lead to a different concept of learnability because randomized PAC learnability is characterized in analogy to simple PAC learnability by the finiteness of the covering number.

Lemma 4.2.1. *If a function class $\mathcal{F}$ with outputs in $\{0,\ldots,B\}$ is randomized PAC learnable with accuracy ϵ and confidence δ on m examples then $\mathcal{F}$ has a finite 2ϵ-covering number.*

Proof. The proof is a direct adaption of [132] (Lemma 6.4). As already mentioned, $N(\epsilon,\mathcal{F},d_P) \leq M(\epsilon,\mathcal{F},d_P)$. Assume that $f_1, \ldots, f_k$ are 2ϵ separated functions and h is a randomized PAC algorithm with accuracy ϵ and confidence δ on m examples. Define $g : \{1,\ldots,k\} \times X^m \times Z^m \times \{0,\ldots,B\}^m \to \{0,1\}$,

$$g(j,\mathbf{x},\mathbf{z},\mathbf{y}) = \begin{cases} 1 & \text{if } d_P(f_j, h_m(\mathbf{y},\mathbf{x},\mathbf{z})) \leq \epsilon, \\ 0 & \text{otherwise}, \end{cases}$$

where $h_m(\mathbf{y},\mathbf{x},\mathbf{z}) = h_m(f,\mathbf{x},\mathbf{z})$ for any function f with values $\mathbf{y}$ on $\mathbf{x}$. For fixed $\mathbf{x}$, $\mathbf{z}$, and $\mathbf{y}$ at most one index j exists where g outputs 1, as a consequence

$$\sum_{\mathbf{y}\in\{0,\ldots,B\}^m} \int_{X^m} \int_{Z^m} \sum_{j=1}^k g(j,\mathbf{x},\mathbf{z},\mathbf{y}) dP_Z^m(\mathbf{z}) dP^m(\mathbf{x}) \leq (B+1)^m .$$

On the other hand, the above term evaluates as

$$\sum_{j=1}^k \int_{X^m} \int_{Z^m} \sum_{\mathbf{y}\in\{0,\ldots,B\}^m} g(j,\mathbf{x},\mathbf{z},\mathbf{y}) dP_Z^m(\mathbf{z}) dP^m(\mathbf{x})$$
$$\geq \sum_{j=1}^k \int_{X^m} \int_{Z^m} g(j,\mathbf{x},\mathbf{z},f(\mathbf{x})) dP_Z^m(\mathbf{z}) dP^m(\mathbf{x}) \geq \sum_{j=1}^k (1-\delta)$$

because of the PAC property. As a consequence, $k \leq (B+1)^m/(1-\delta)$. □

The purpose of the following definition is to deal with training data which are corrupted by some noise.

Definition 4.2.3. *A function class $\mathcal{F}$ is PAC learnable with accuracy ϵ and confidence δ on m examples with noise from a set of probability distributions $\mathcal{D}$ on X with variance σ^2 if a learning algorithm h exists such that*

$$\sup_{f\in\mathcal{F}} \sup_{D\in\mathcal{D},\text{var}(D) = \sigma^2} P^m \times D^m((\mathbf{x},\mathbf{n}) \mid d_P(h_m(f+\mathbf{n},\mathbf{x}),f) > \epsilon) \leq \delta ,$$

where $h_m(f+\mathbf{n},\mathbf{x})$ is a short notation for $h_m(x_1, f(x_1)+n_1, \ldots, x_m, f(x_m)+n_m)$.

Obviously, learnability of a function class with data which is corrupted by some noise according to a distribution D corresponds to learning exact data with a randomized algorithm which first adds some noise to the data. But one can even show that learning noisy data is almost the same as learning exact data in an appropriately quantized class with a finite output set if the noise fulfills some regularity conditions: The regularity conditions are fulfilled, for example, for Gaussian or uniform noise and demand that the distributions have zero mean, finite variance, and are absolutely continuous, and the corresponding densities have a total variation, which can be bounded in dependence on the variance. In this case, D is called *admissible* for short.

Under these conditions the following has been shown in [7]: Any PAC algorithm which learns $\mathcal{F}$ with accuracy ϵ and confidence δ on m examples with noisy data gives rise to a probabilistic PAC algorithm which learns a quantized version $\mathcal{F}_\alpha$ of $\mathcal{F}$ with accuracy 2ϵ and confidence 2δ on m examples. The noise of the algorithm depends on the noise in D. $\mathcal{F}_\alpha$ is the class with outputs in $\{0, \alpha, 2\alpha, \ldots, 1\}$, which is obtained if all outputs of $f \in \mathcal{F}$ in $[k\alpha - \alpha/2, k\alpha + \alpha/2[$ are identified with $k\alpha$. α depends on ϵ, δ, D, and the number of examples. Lemma 5 in [7] is stated for distribution-independent PAC learnability. But the proof is carried out for any specific distribution P as well.

Now from Lemma 4.2.1 it follows that $\mathcal{F}_\alpha$ is randomized PAC learnable only if it has a finite covering number. Obviously, this leads to a finite covering number of $\mathcal{F}$, too, since the distance of any function to its quantized version is at most $\alpha/2$. As a consequence, PAC learnability in the distribution-dependent setting under realistic conditions, *i.e.*, the presence of noise, implies the finiteness of the covering number.

It may happen that we do not know whether the data is noisy, but in any case we are aware of the special form of our algorithm, for example, it may be PUAC and stable. Stability of an algorithm allows some kind of disturbances of the data, and it can indeed be shown that any stable PUAC algorithm learns $\mathcal{F}$ with noisy data, too, if the variance of the noise is limited in some way. The probability distributions D have *light tails* if positive constants c_0 and s_0 exist such that $D(n \mid |n| \geq s/2) \leq c_0 e^{-s/\sigma}$ for all distributions $D \in \mathcal{D}$ with variance σ^2 and $s > s_0\sigma$. This is fulfilled for Gaussian noise, for example.

Theorem 4.2.5. *Assume that $\mathcal{F}$ is learnable with a PUAC algorithm with accuracy ϵ_2 and confidence δ on m examples. Assume that the algorithm is ϵ_1-ϵ_2-stable with confidence δ on m examples. Then $\mathcal{F}$ is PUAC learnable with accuracy $3\epsilon_2$ and confidence 3δ on m examples with noise from a class $\mathcal{D}$ with light tails and variance σ^2 bounded according to the confidence δ.*

Proof. We are given $D \in \mathcal{D}$ with variance σ^2 and h as stated in the theorem, define a learning algorithm $\bar{h}$ on the noisy data as follows: Given a sample $(x_i, f(x_i) + n_i)_i$ the algorithm chooses a sample $(x_i, f_1(x_i))_i$ such that $\hat{d}_m(f + \mathbf{n}, f_1, \mathbf{x})$ is minimal for $f_1 \in \mathcal{F}$ and takes this exact sample as an input for h. For $\bar{h}$ the following inequality is valid:

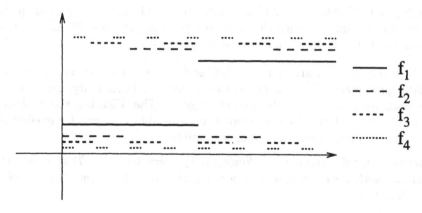

Fig. 4.1. Function class with infinite covering number which is PUAC learnable

$$P^m \times D^m((\mathbf{x}, \mathbf{n}) \mid \sup_f d_P(\bar{h}_m(f + \mathbf{n}, \mathbf{x}), f) > 3\epsilon_2)$$
$$\leq \quad D^m(\mathbf{n} \mid \textstyle\sum_i |n_i| \geq m\epsilon_1/2) + P^m(\mathbf{x} \mid \exists f \, d_P(h_m(f, \mathbf{x}), f) > \epsilon_2)$$
$$+ P^m(\mathbf{x} \mid \exists f, g \, (\hat{d}_m(f, g, \mathbf{x}) \leq \epsilon_1 \wedge d_P(h_m(f, \mathbf{x}), h_m(g, \mathbf{x})) > \epsilon_2)),$$

where the second probability is bounded because of the PUAC property, the
third probability is bounded due to the stability of h, and the first probability
is bounded for distributions with light tails and sufficiently small variance as
follows:

$$D^m(\textstyle\sum_i |n_i| \geq m\epsilon_1/2)$$
$$\leq \quad \mathrm{E}_{D^m}(4(\textstyle\sum |n_i|)^2)/(m^2 \epsilon_1^2)$$
$$= \quad 4/\epsilon_1^2 \cdot \left(\mathrm{Var}(|n_1|)/m + \mathrm{E}(|n_1|)^2 \right)$$
$$\leq \quad 4\sigma^2/(\epsilon_1^2 m) + 4/\epsilon_1^2 \left(s_0\sigma/2 + \int_{s_0\sigma/2}^{\infty} c_0 e^{-2x/\sigma} dx \right)^2$$
$$= \quad \sigma^2/\epsilon_1^2 \left(4/m + (s_0 + c_0 e^{-s_0})^2 \right).$$

$\square$

This result implies that the covering number of a function class is finite if
the class is ϵ-consistently PUAC learnable. The demand for some robustness
of the learning algorithm prohibits an exact encoding of the functions in the
output space. On the contrary, learnability at the PUAC level may only be
due to the possibility of encoding functions in real values. One example of
this method is the following function class:

Example 4.2.6. A function class exists which is PUAC learnable and has an
infinite covering number.

Proof. Consider the function class $\{f : [0, 1] \to [0, 1] \mid \exists l \geq 1 \, f = f_l\}$, where

$$f_l(x) = \begin{cases} 1 - (1 + e^{-l})^{-1} & \text{if } x \in \bigcup_{i=0}^{2^{l-1}-1}[2i/2^l, (2i+1)/2^l], \\ (1 + e^{-l})^{-1} & \text{otherwise} \end{cases}$$

(see Fig. 4.1). Consider the uniform distribution. The output values uniquely determine the function such that any consistent algorithm is PUAC but two functions of the class have a distance of at least 0.2.   □

Now the question arises as to whether finiteness of the covering number is still guaranteed if we do not consider uniform learnability, and only a tolerance with respect to the inputs is given. The following result shows that even at the level of ϵ-consistent PAC learnability an exact encoding of functions into real values is no longer possible.

Theorem 4.2.6. *Assume $\mathcal{F}$ is finitely ϵ_1-ϵ_2-characterizable. Then $\mathcal{F}$ is PAC learnable with accuracy ϵ_2 and any noise D, where the support is bounded according to ϵ_1.*

Proof. If $\mathcal{F}$ is finitely characterizable, any probabilistic algorithm with small empirical error is PAC, too, because

$$\sup_{f \in \mathcal{F}} P^m \times P_Z^m((\mathbf{x}, \mathbf{z}) \mid d_P(f, h_m(f, \mathbf{x}, \mathbf{z})) > \epsilon_2)$$
$$\leq \quad \sup_{f \in \mathcal{F}} P^m \times P_Z^m((\mathbf{x}, \mathbf{z}) \mid \hat{d}_m(f, h_m(f, \mathbf{x}, \mathbf{z}), \mathbf{x}) > \epsilon_1)$$
$$+ \sup_{f \in \mathcal{F}} P^m(\mathbf{x} \mid \exists g \, (\hat{d}_m(f, g, \mathbf{x}) \leq \epsilon_1 \wedge d_P(f, g) > \epsilon_2)).$$

As already mentioned, learning on noisy data with an algorithm h is the same as learning on exact data with a probabilistic algorithm which first adds some noise to the data and afterwards applies h. If the support of the noise is bounded such that the probability of events with elements contained in $\{n \mid |n| > \epsilon_1/2\}$ is 0, then an algorithm h on the noisy data, choosing one function with smallest empirical distance of the training data, gives rise to an ϵ_1-consistent probabilistic learning algorithm on the exact data. This algorithm as well as the original one is PAC.   □

4.2.4 Model-free Learning

Up to now, all results in this chapter have been stated for the model-dependent case where the form of the function that are to be learned is well known and the function itself is contained in $\mathcal{F}$.

If the function that is to be learned is not contained in $\mathcal{F}$ and the special form is unknown, still most of the previous notations and questions make sense in the model-free case. The terms PAC and PUAC have already been expanded to this case, the generalization of the terms 'consistent', 'finitely characterizable', ... to the model-free case are immediate: From all empirical distances we subtract the minimum empirical error $\hat{J}_m(f, \mathbf{x})$, from all real distances we subtract the minimum real distance $J_m(f)$, where f is the function that has to be learned. The only thing we have to take care of is that all functions denoted by f in the definitions are from the class $\mathcal{F}_0$ of functions that has to be learned, and the functions denoted by g, the approximating functions, are elements of $\mathcal{F}$.

If we scan the theorems of this paragraph it turns out that the proofs which relate PUAC learnability to consistent PUAC learnability no longer hold in the model-free case. They rely on the symmetric role of g and f, which is no longer guaranteed if these functions are taken from different classes. But the argumentation which shows the equivalence of ϵ_1-consistently PAC or PUAC learnability with accuracy ϵ_2 and finitely ϵ_1-ϵ_2-characterizability or the scaled shrinking width property transfer to the model-free case – provided that we can verify that any partial mapping we constructed in the proofs can be completed to a learning algorithm with several properties. In the special situations this condition reads as the existence of a consistent or asymptotically ϵ_1-consistent learning algorithm. The existence of a consistent algorithm is not guaranteed for a function class where the term $\hat{J}_m(f, \mathbf{x})$ is an infimum and not a minimum. Therefore the guarantee of any consistent algorithm being PAC or PUAC does not imply anything in this case. But here, the scale sensitive versions of the above terms turn out to be useful because obviously a nearly consistent learning algorithm can be found in any case.

4.2.5 Dealing with Infinite Capacity

We conclude the discussion of learnability with a generalization of the two approaches to function classes that allow a stratification of the learning problem, as mentioned in the previous section. First of all, we consider a class of functions $\mathcal{F}$ with inputs in $X = \bigcup_{i=0}^{\infty} X_i$ where $X_i \subset X_{i+1}$ for the subspaces X_i. Denote the probability of X_i by p_i and the probability measure which is induced on X_i by P_i. Denote the restriction of $\mathcal{F}$ to inputs X_i by $\mathcal{F}_i$ and the restriction of one single function by $f|X_i$. First of all, $\mathcal{F}$ is PAC learnable if and, in the case of noisy data, only if the covering number of $\mathcal{F}$ is finite. A useful criterion to test this finiteness is to bound the capacity of the single classes $\mathcal{F}_i$. This method uses the natural stratification of $\mathcal{F}$ via the input set and leads to a weaker condition than the finiteness of the VC- or pseudo-dimension of the whole class $\mathcal{F}$. One can derive explicit bounds as follows:

Theorem 4.2.7. *Take i such that $p_i > 1-\epsilon$. Assume that $\mathcal{VC}(\mathcal{F}_i)$ or $\mathcal{PS}(\mathcal{F}_i)$ is finite. Then $\mathcal{F}$ is PAC learnable with accuracy ϵ and confidence δ with an algorithm which uses*

$$\frac{32}{\epsilon} \left(\ln \frac{4}{\delta} + \mathcal{VC}(\mathcal{F}_i) \ln \left(\frac{4ep_i}{\epsilon + p_i - 1} \ln \left(\frac{4ep_i}{\epsilon + p_i - 1} \right) \right) \right)$$

examples if $\mathcal{F}$ is a concept class and

$$\frac{8}{\epsilon^2} \left(\ln \frac{2}{\delta} + \mathcal{PS}(\mathcal{F}_i) \ln \left(\frac{4ep_i}{\epsilon + p_i - 1} \ln \left(\frac{4ep_i}{\epsilon + p_i - 1} \right) \right) \right)$$

examples if $\mathcal{F}$ is a function class. This is polynomial in $1/\epsilon$ and $1/\delta$ if $p(X \backslash X_i) \leq d_i^{-\beta}$ for $d_i = VC(\mathcal{F}_i)$ in the concept case and $d_i = PS(\mathcal{F}_i)$ in the function case and some positive β.

Proof. For $f, g \in \mathcal{F}$ we find $d_P(f, g) \leq (1 - p_i) + p_i d_{P_i}(f|X_i, g|X_i)$, therefore

$$M(\epsilon, \mathcal{F}, d_P) \leq M\left(\frac{\epsilon + p_i - 1}{p_i}, \mathcal{F}_i, d_{P_i}\right)$$

for $\epsilon > 1 - p_i$. This term is bounded by

$$2\left(\frac{2ep_i}{\epsilon + p_i - 1} \ln\left(\frac{2ep_i}{\epsilon + p_i - 1}\right)\right)^{d_i}$$

with $d_i = VC(\mathcal{F}_i)$ or $d_i = PS(\mathcal{F}_i)$, respectively. The minimum risk algorithm is PAC and takes the number of examples as stated above [132] (Theorems 6.3 and 6.4). These numbers are polynomial in $1/\epsilon$ and $1/\delta$ if d_i is polynomial, too, which is fulfilled if $p_i > 1 - (d_i)^{-\beta}$ for some $\beta > 0$ because this inequality implies $p_i > 1 - \epsilon$ for all $d_i > \epsilon^{-1/\beta}$.   □

If we want to use an arbitrary algorithm which minimizes the empirical error instead of the minimum risk algorithm we can use the following result as a guarantee for the generalization ability.

Theorem 4.2.8. *Any function class $\mathcal{F}$, where all $\mathcal{F}_i$ have finite VC- or fat shattering dimension, respectively, has the UCEM property.*

Proof. The empirical covering number of $\mathcal{F}$ can be bounded as follows:

$$
\begin{aligned}
&E_{P^m}(\lg N(\epsilon, \mathcal{F}|\mathbf{x}, \hat{d}_m))/m \\
\leq\ &(P(\text{at most } m(1 - \epsilon/2) \text{ points are contained in } X_i) \lg(2/\epsilon)^m \\
&+ E_{P_i^m}(\lg M(\epsilon/2, \mathcal{F}_i|\mathbf{x}, \hat{d}_m)))/m \\
\leq\ &\lg(2/\epsilon) \cdot 16p_i(1 - p_i)/(m\epsilon^2) + E_{P_i^m}(\lg N(\epsilon/4, \mathcal{F}_i|\mathbf{x}, \hat{d}_m))/m
\end{aligned}
$$

for $p_i \geq 1 - \epsilon/4$ due to the Tschebyschev inequality. For a concept class the last term in the sum is bounded by

$$\frac{1}{m}\left(1 + VC(\mathcal{F}_i) \lg\left(\frac{8e}{\epsilon} \ln\left(\frac{8e}{\epsilon}\right)\right)\right).$$

For a function class an upper bound is

$$\frac{1}{m}\left(1 + d_i \ln\left(\frac{8em}{d_i\epsilon} \cdot \lg\left(\frac{4 \cdot 16m}{\epsilon^2}\right)\right)\right)$$

with $d_i = fat_{\epsilon/16}(\mathcal{F}_i)$. Obviously, these terms tend to 0 for $m \to \infty$.   □

These arguments ensure that with some prior knowledge about the probability distribution even the UCEM property of $\mathcal{F}$ can be guaranteed if only the capacity of the single subclasses $\mathcal{F}_i$ is restricted.

In contrast, the luckiness framework allows us to introduce an arbitrary hierarchy which may even depend on the actual training data. Furthermore, no prior knowledge about the membership in one subclass of the hierarchy is needed. The fact that the learning scenario belongs to a certain subclass can be decided a posteriori and is substituted only by some confidence about the expected difficulty of the learning task.

Assume that $\mathcal{F}$ is a $[0, 1]$-valued function class with inputs in X and

$$L : X^m \times \mathcal{F} \to \mathbb{R}^+$$

is a function, the so-called *luckiness function*. This function measures some quantity, which allows a stratification of the entire function class into subclasses with some finite capacity. If L outputs a small value, we are lucky in the sense that the concrete output function of a learning algorithm is contained in a subclass of small capacity which needs only few examples for correct generalization. Define the corresponding function

$$l : X^m \times \mathcal{F} \times \mathbb{R}^+ \to \mathbb{R}^+, \quad l(\mathbf{x}, f, \alpha) = |\{g|\mathbf{x} \mid g \in \mathcal{F}, L(\mathbf{x}, g) \geq L(\mathbf{x}, f)\}_\alpha|,$$

which measures the number of functions that are at least as lucky as f on $\mathbf{x}$. Note that we are dealing with real outputs, consequently a quantization according to some value α of the outputs is necessary to ensure the finiteness of the above number: As defined earlier, $\mathcal{F}_\alpha$ refers to the function class with outputs in $\{0, \alpha, 2\alpha, \ldots\}$, which is obtained if all outputs of $f \in \mathcal{F}$ in $[k\alpha - \alpha/2, k\alpha + \alpha/2[$ are identified with $k\alpha$ for $k \in \mathbb{N}$. The luckiness function L is *smooth* with respect to η and Φ, which are both mappings $\mathbb{N} \times (\mathbb{R}^+)^3 \to \mathbb{R}^+$ if

$$P^{2m}(\mathbf{xy} \mid \exists g \in \mathcal{F} \, \forall \mathbf{x'y'} \subset_\eta \mathbf{xy} \, l(\mathbf{x'y'}, g, \alpha) > \Phi(m, L(\mathbf{x}, g), \delta, \alpha)) \leq \delta,$$

where $\mathbf{x'y'} \subset_\eta \mathbf{xy}$ indicates that a fraction $\eta = \eta(m, L(\mathbf{x}, g), \delta, \alpha)$ is deleted in $\mathbf{x}$ and in $\mathbf{y}$ to obtain $\mathbf{x'}$ and $\mathbf{y'}$. This is a stronger condition than the smoothness requirement in [113] because the consideration is not restricted to functions g that coincide on $\mathbf{x}$. Since we want to get results for learning algorithms with small empirical error, but which are not necessarily consistent, this generalized possibility of estimating the luckiness of a double sample knowing only the first half is appropriate in our case.

Now in analogy to [113] we can state the following theorem which guarantees some kind of UCEM property if the situation has turned out to be lucky in a concrete learning task.

Theorem 4.2.9. *Suppose p_i ($i \in \mathbb{N}$) are positive numbers with $\sum_i p_i = 1$, and L is a luckiness function for a class $\mathcal{F}$ which is smooth with respect to η and Φ. Then the inequality*

$$\inf_f \inf{} f \in \mathcal{F} P^m(\mathbf{x} \mid \forall i \, (\Phi(m, L(\mathbf{x}, h_m(f, \mathbf{x})), \delta, \alpha) \leq 2^{i+1}$$
$$\Rightarrow |\hat{d}_m(f, h_m(f, \mathbf{x}), \mathbf{x}) - d_P(f, h_m(f, \mathbf{x}))| \leq \epsilon(m, i, \delta, \alpha))) \geq 1 - \delta$$

is valid for any learning algorithm h, real values δ, $\alpha > 0$, and

$$\epsilon(m, i, \delta, \alpha) = 4\alpha + 8\eta + 4\sqrt{2\eta \ln(m) + \frac{1}{m}\left((i+3)\ln 2 + \ln \frac{4}{p_i \delta}\right)}$$

where $\eta = \eta(m, L(\mathbf{x}, h_m(f, \mathbf{x})), (p_i\delta)/4, \alpha)$.

Proof. For any $f \in \mathcal{F}$ we can bound the probability

$$P^m(\mathbf{x} \mid \exists h \exists i \, (\Phi(m, L(\mathbf{x}, h), \delta, \alpha) \leq 2^{i+1} \wedge |\hat{d}_m(f, h, \mathbf{x}) - d_P(f, h)| > \epsilon))$$
$$\leq \; 2P^{2m}(\mathbf{xy} \mid \exists h \exists i \, (\Phi(m, L(\mathbf{x}, h), \delta, \alpha) \leq 2^{i+1}$$
$$\wedge |\hat{d}_m(f, h, \mathbf{x}) - \hat{d}_m(f, h, \mathbf{y})| > \epsilon/2))$$

for $m \geq 2/\epsilon^2$ which is fulfilled for ϵ as defined above [132] (Theorem 5.7, step 1). It is sufficient to bound the probability of the latter set for each single i by $(p_i\delta)/2$. Intersecting such a set for a single i with a set that occurs at the definition of the smoothness of l and its complement, respectively, we obtain the bound

$$2P^{2m}(\mathbf{xy} \mid \exists \mathbf{x}'\mathbf{y}' \subset_\eta \mathbf{xy} \, \exists h$$
$$(l(\mathbf{x}'\mathbf{y}', h, \alpha) \leq 2^{i+1} \wedge |\hat{d}_m(f, h, \mathbf{x}) - \hat{d}_m(f, h, \mathbf{y})| > \epsilon/2)) + \frac{p_i\delta}{4}$$

where $\eta = \eta(m, L(\mathbf{x}, h), (p_i\delta)/4, \alpha)$. Denote the above event by A. Consider the uniform distribution U on the group of permutations in $\{1, \ldots, 2m\}$ that only swap elements j and $j+m$ for some j. It can be found that

$$P^{2m}(\mathbf{xy} \mid \mathbf{xy} \in A)$$
$$= \; \int_{X^{2m}} U(\sigma \mid (\mathbf{xy})^\sigma \in A) dP^{2m}(\mathbf{xy})$$
$$\leq \; \sup_{\mathbf{xy}} U(\sigma \mid (\mathbf{xy})^\sigma \in A),$$

where $\mathbf{x}^\sigma$ is the vector obtained by applying the permutation σ [132] (Theorem 5.7, step 2). The latter probability can be bounded by

$$\sup_{\mathbf{xy}} \sum_{\mathbf{x}'\mathbf{y}' \subset_\eta \mathbf{xy}} U(\sigma \mid \exists h \, (l((\mathbf{x}'\mathbf{y}')^\sigma, h, \alpha) \leq 2^{i+1} \wedge$$
$$|m'\hat{d}_{m'}(f_\alpha, h_\alpha, \mathbf{x}') - m'\hat{d}_{m'}(f_\alpha, h_\alpha, \mathbf{y}')| > m(\epsilon/2 - 2\alpha - 2\eta))),$$

where $m' = m(1 - \eta)$ is the length of $\mathbf{x}'$ and $\mathbf{y}'$. f_α denotes the quantized version of f, where outputs in $[k\alpha - \alpha/2, k\alpha + \alpha/2[$ are identified with $k\alpha$. Now denote the event of which the probability U is measured by B, and define equivalence classes C on the permutations such that two permutations belong to the same class if they map all indices to the same values unless $\mathbf{x}'$ and $\mathbf{y}'$ both contain this index. We find that

$$U(\sigma \mid (\mathbf{x}'\mathbf{y}')^\sigma \in B) \;=\; \sum_C P(C) U(\{\sigma \mid (\mathbf{x}'\mathbf{y}')^\sigma \in B\} \mid C)$$
$$\leq \; \sup_C U(\{\sigma \mid (\mathbf{x}'\mathbf{y}')^\sigma \in B\} \mid C).$$

If we restrict the events to C we definitely consider only permutations which swap elements in $\mathbf{x}'$ and $\mathbf{y}'$ such that we can bound the latter probability by

$$2^{i+1} \sup_h U'(\sigma \mid |m'\hat{d}_{m'}(f_\alpha, h_\alpha, (\mathbf{x}')^\sigma) - m'\hat{d}_{m'}(f_\alpha, h_\alpha, (\mathbf{y}')^\sigma)| > m(\epsilon/2 - 2\alpha - 2\eta))$$

where U' denotes the uniform distribution on the swappings of the common indices of $\mathbf{x}'$ and $\mathbf{y}'$. The latter probability can be bounded using Hoeffding's inequality for random variables $\mathbf{z}$ with values in $\{\pm(\text{error on } x'_i - \text{error on } y'_i)\}$ by the term

$$2e^{-m(\epsilon/2 - 2\alpha - 4\eta)^2/(2(1-\eta))}.$$

In total, we can therefore obtain the desired bound if we choose ϵ such that

$$\frac{p_i\delta}{4} \geq 2\left(\frac{m}{\eta m}\right)^2 2^{i+1} 2e^{-m(\epsilon/2 - 2\alpha - 4\eta)^2/(2(1-\eta))},$$

which is fulfilled for

$$\epsilon \geq 4\alpha + 8\eta + 4\sqrt{1-\eta} \cdot \sqrt{\frac{(i+3)\ln 2}{m} + 2\eta \ln(m) + \frac{\ln(4/(p_i\delta))}{m}}.$$

$\square$

Note that the bound ϵ tends to 0 for $m \to \infty$ if α and η are decreasing, η in such a way that $\eta \ln m$ becomes small. Furthermore, we have obtained bounds on the difference between the real and empirical error instead of dealing only with consistent algorithms as in [113]. We have considered functions instead of concept classes which cause an increase in the bound by α due to the quantization, and a decrease in the convergence because we have used Hoeffding's inequality in the function case.

Furthermore, a dual formulation with an unluckiness function L' is possible, too. This corresponds to a substitution of $\leq$ by $\geq$ in the definition of l; the other formulas hold in the same manner.

4.3 Bounds on the VC-Dimension of Folding Networks

The VC- or fat shattering dimension plays a key role for the concrete computation of the various bounds, hence we first want to estimate these dimensions for folding networks.

4.3.1 Technical Details

At the beginning we prove some lemmata showing that several restrictions of the network architecture do not alter the dimension by more than a constant factor.

Lemma 4.3.1. *Let $h \circ \tilde{g}_y$ be a recursive architecture where h and g are multilayered feed-forward architectures. Then an architecture $\tilde{f}_{y'}$ exists with only one layer and the same number of weights and computation neurons such that it has the same behavior as the original one if a time delay according to the number of layers is introduced to the inputs.*

Proof. The idea of the proof is simple: Instead of computing the activation of several layers in a multilayer feed-forward architecture in one time step we compute the activation in several recursive steps such that the activation of only one layer is to be computed at each time step (see Fig. 4.2 as an example).

Denote the number of neurons in the single computation layers by $n + k \cdot l$, $i_1, \ldots, i_{h_g}, l, i'_1, \ldots, i'_{h_h}, m$, where h_h is the number of hidden layers in h, and h_g is the number of hidden layers in g. Instead of previously $k \cdot l$ context neurons we now take $k \cdot (l + i_1 + \ldots + i_{h_g} + i'_1 + \ldots + i'_{h_h} + m)$ context neurons such that a corresponding neuron in the context layer of $\tilde{f}_{y'}$ exists for each computation neuron in $h \circ \tilde{g}_y$. The connections in $\tilde{f}_{y'}$ directly correspond to the connections in $h \circ \tilde{g}_y$. To be more precise, denote the context neurons in the input layer of $\tilde{f}_{y'}$ by n_i^j ($i = 1, \ldots, 1 + i_1 + \ldots + m$, $j = 1, \ldots, k$, one can find k sets of context neurons according to the fan-out of the input trees), and the corresponding neurons in the computational part by n'_i. Then the connections of the k sets of neurons $n_1^j, \ldots, n_l^j$ and of the input neurons of the entire architecture to the neurons $n'_{l+1}, \ldots, n'_{l+i_1}$ correspond to the connections in the first layer of g, the connections between $n^1_{l+1}, \ldots, n^1_{l+i_1}$ to n'_{l+i_1+1}, $\ldots, n'_{l+i_1+i_2}$ correspond to the connections in the second layer of g, analogous for the layers 3 to h_g in g, $n^1_{l+i_1+\ldots+i_{h_g}-1+1}, \ldots, n^1_{l+i_1+\ldots+i_{h_g}}$ are connected to $n'_1, \ldots, n'_l$ corresponding to the last layer in g, $n^1_1, \ldots, n^1_l$ are connected to $n'_{l+i_1+\ldots+i_{h_g}+1}, \ldots, n'_{l+i_1+\ldots+i_{h_g}+i'_1}$ corresponding to layer 1 in h, $n^1_{l+i_1+\ldots+i_{h_g}+1}, \ldots, n^1_{l+i_1+\ldots+i_{h_g}+i'_1}$ are connected to $n'_{l+i_1+\ldots+i_{h_g}+i'_1+1}, \ldots,$ $n'_{l+i_1+\ldots+i_{h_g}+i'_1+i'_2}$ corresponding to layer 2 in h, analogous for the layers 3, $\ldots, h_h$ in h, $n^1_{l+i_1+\ldots+i_{h_g}+i'_1+\ldots+i'_{h_h}-1+1}, \ldots, n^1_{l+i_1+\ldots+i_{h_g}+i'_1+\ldots+i'_{h_h}}$ are connected to the outputs $n'_{l+i_1+\ldots+i_{h_g}+i'_1+\ldots+i'_{h_h}+1}, \ldots, n^1_{l+i_1+\ldots+i_{h_g}+i'_1+\ldots+i'_{h_h}+m}$ corresponding to the last layer in h. All other connections are 0. See Fig. 4.2 as an example.

Obviously, the number of computation neurons and connections are identical in $\tilde{f}_{y'}$ and $h \circ \tilde{g}_y$. Furthermore, $\tilde{f}_{y'}$ has the same behavior as the original function, provided that the concrete weight values are the same, $y' = (y, 0, \ldots, 0)$, and a time delay is introduced to the inputs. This means that any subtree $a(t_1, \ldots, t_k)$ in an input tree is recursively substituted by $\underbrace{a(\ldots a(t_1, \ldots, t_k) \ldots)}_{h_g}$ and afterwards, the entire tree $a(t_1, \ldots, t_k)$ is substi-

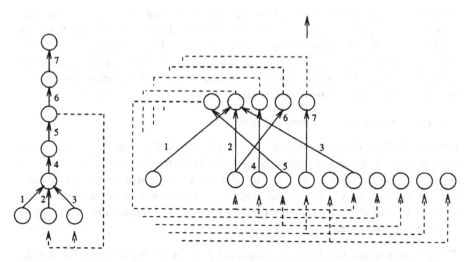

Fig. 4.2. Simulating hidden layer in a folding architecture: Instead of the neurons in the single layers, several new context neurons are introduced; the links from the original network connect the context neurons.

tuted by $\underbrace{a(\ldots a(t_1,\ldots,t_k)\ldots)}_{h_h+1}$. The delay in the single subtrees just enables

us to perform one computation step in g via h_g recursive computation steps in $\tilde{f}_{y'}$, the delay at the root enables us to perform the final computation in h via $h_h + 1$ recursive steps in $\tilde{f}_{y'}$.   □

As a consequence of this argumentation a restriction of the architecture to one layer only leads to a decrease in the various dimensions by a constant factor compared to a multilayered architecture. Due to the time delay the input length has to be substituted by a multiple of this value. Note that this argumentation transfers to arbitrary architectures, too, if the activation function is capable of approximating the identity in an appropriate way. Then an architecture with arbitrary connection structure can be approximated on each finite set by a layered architecture, which can be simulated as described above.

A second observation is that we can drop all biases, too.

Lemma 4.3.2. *Given a bias vector θ, any architecture $h \circ \tilde{g}_y$ can be simulated by an architecture with bias vector $\mathbf{0}$. This requires one additional input neuron which receives constant input 1, one additional context neuron, and an additional number of weights corresponding to the biases.*

Proof. It is well known that biases in feed-forward architectures can be simulated with a connection to an additional input neuron which receives constant input $x \in \mathbb{R}$ different from 0. Just choose the additional weight w_i of neuron

i as $(\theta_i' - \theta_i)/x$ for any bias θ_i'. We can proceed in the same way to simulate biases in the recursive part $\tilde{g}_y$. However, since direct connections from an input neuron to neurons in h are not allowed by definition we have to add one additional output neuron to g, whose only predecessor is the additional input neuron. This neuron stores a constant value and can play the role of an additional constant neuron for part h to simulate the biases in h.   □

The same is valid for the initial context:

Lemma 4.3.3. *Given an initial context* y', *any architecture* $h \circ \tilde{g}_y$ *where the components of the initial context* y *are restricted to elements of the range of the activation function of the outputs of* g *can be simulated by an architecture with initial context* y'. *The simulating architecture requires an additional input neuron, additional weights according to* y, *and an expansion of the input trees by one time step.*

Proof. We expand the input dimension by 1. In all input trees for $h \circ \tilde{g}_y$ the input labels are expanded by one component 0 for the additional input neuron. Furthermore, all leaves a are substituted by $a(0, \ldots, 0, 1)$. This additional time step just writes the images of the additional weights plus a term depending on the bias into the context neurons. If the components of y are contained in the range of the activation function the additional weights can be chosen such that this first computed context vector coincides with y.   □

This argumentation enables us to expand any input tree to an equivalent tree with full range of some fixed height t. Here full range means that in levels 1 to t all labels are filled with vectors and empty subtrees can only be found at the level $t + 1$. We will use this fact to argue that we can restrict the VC-analysis of recurrent architectures and inputs with restricted height to the analysis of inputs of one fixed structure, and therefore the analysis of one feed-forward architecture which is obtained by unfolding.

Lemma 4.3.4. *Any architecture* $h \circ \tilde{g}_y$ *can be modified to an architecture with the same behavior such that any input of height at most* t *can be expanded to an equivalent input of height exactly* t *with full range. The number of weights and neurons is of the same order.*

Proof. First, we assume that the activation fulfills $\sigma(0) = 0$. We can modify the architecture such that the biases and the initial context are 0. Then any expansion of an input tree by subtrees with labels 0 does not change the output of the tree.

If $\sigma(0) \neq 0$ we fix the biases of the neurons in the previous architecture to $-p\sigma(0)$, where p is the number of predecessors of the corresponding neuron, not counting the input neurons of the architecture, and we fix the initial context to 0. Then we expand any input tree with entries 0 to a tree of full range. The bias corresponds to a translation of the activation function such that 0 is mapped to 0, with the only difficulty that at the first computation step the neurons in the first layer receive the additional activation $-p\sigma(0)$ instead of 0. This difficulty can be overcome by introducing one additional input unit with a constant weight $p\sigma(0)$ to all neurons in the first layer. This input receives an activation 1 at the leaves of the expanded input tree and an activation 0 at all other time steps. □

It follows from these technical lemmata that several restrictions of the architecture do not affect the VC-dimension by more than a constant factor. Furthermore, the last argumentation enables us to restrict the consideration to inputs of a fixed structure if we want to estimate the VC-dimension of an architecture with inputs of a limited height.

The possibility of comparing the outputs with arbitrary values instead of 0 in the definition of the pseudo-dimension causes some difficulties in obtaining upper bounds and can in fact simply be dropped because of the equality $PS(\mathcal{F}) = VC(\mathcal{F}_e)$, where $\mathcal{F}_e = \{f_e : X \times \mathbb{R} \to \{0,1\} \mid f_e(x,y) = H(f(x) - y)$ for some $f \in \mathcal{F}\}$ [86]. At least for standard architectures the consideration of $\mathcal{F}_e$ instead of $\mathcal{F}$ only causes a slight constant increase in the number of parameters.

For a concept class which is parameterized by a set Y, that means $\mathcal{F} = \{f_y : X \to \{0,1\} \mid y \in Y\}$, we can define the dual class $\mathcal{F}^\vee = \{f.(x) : Y \to \{0,1\}, (f.(x))(y) = f_y(x) \mid x \in X\}$. It is well known that $VC(\mathcal{F}) \geq \lfloor \lg VC(\mathcal{F}^\vee) \rfloor$ [76]. Since neural architectures are parameterized by the weights, this fact is a useful argument to derive bounds for the VC-dimension of neural architectures. The same argumentation as in [76] shows that an analogous inequality is valid for uniform versions of the pseudo-dimension and fat shattering dimension, too. Here uniform means that in the definition of the fat shattering dimension all reference points for the classification are required to be equal for the points that are shattered. Obviously, this can only decrease the fat shattering dimension. The inequality for the fat shattering dimension is not valid for the nonuniform version, as can be seen in the following example.

Example 4.3.1. A function class $\mathcal{F}$ exists with $\lg \lfloor fat_\alpha(\mathcal{F}) \rfloor > fat_\alpha(\mathcal{F}^\vee)$.

Proof. Consider the sets $X = \{i/n \mid i = -n + 1, \ldots, n - 1\}$ and $Y = \{-1, 1\}^{2n-1}$ and function $F : X \times Y \to [-1, 1], (i/n, \epsilon_1, \ldots, \epsilon_{2n-1}) \mapsto (2i + \epsilon_i)/(2n)$. F induces the classes $\mathcal{F} = \{F_y : X \to [-1, 1] \mid y \in Y\}$ with $fat_\alpha(\mathcal{F}) = 2n - 1$ and $\mathcal{F}^\vee = \{F_x : Y \to [-1, 1] \mid x \in X\}$ with $fat_\alpha(\mathcal{F}^\vee) = 1$ for $\alpha < 1/(2n)$. □

4.3.2 Estimation of the VC-Dimension

We first derive upper bounds for the various dimensions which measure the capacity of recurrent and folding architectures. In a first step all inputs are expanded to input trees with full range and height exactly t to obtain upper bounds on the VC- or pseudo-dimension of folding architectures with N neurons, W weights including biases, and inputs of height at most t. Afterwards, the dimensions of the corresponding feed-forward architectures obtained via unfolding are limited. This unfolding technique has already been applied by Koiran and Sontag to simple recurrent architectures [70]. For folding architectures with a fan-out $k \geq 2$ and activation function σ the following bounds can be obtained for $d = \mathcal{VC}(\mathcal{F}|X_t)$ or $\mathcal{PS}(\mathcal{F}|X_t)$, respectively, if X_t denotes the set of input trees of height at most t:

$$d = \begin{cases} O(k^t W \ln(k^t W)) & \text{if } \sigma = \text{H}, \\ O(W^2 k^{2t} N^2) & \text{if } \sigma = \text{sgd}, \\ O(W N k^t \ln q + W th \ln d) & \text{if } \sigma \text{ is piecewise polynomial}, \\ & q \text{ is the maximum number of pieces}, \\ & d \text{ is the maximum degree}, \\ & \text{and } h \text{ is the depth of the network}, \\ O(W \ln(th)) & \text{if } \sigma \text{ is linear}, h \text{ denotes the depth}. \end{cases}$$

For large t the bound in the case $\sigma = \text{H}$ can be improved with the same techniques as in [70].

Theorem 4.3.1. $\mathcal{VC}(\mathcal{F}|X_t) = O(NWk + Wt \ln k + W \ln W)$ if $\sigma = \text{H}, k \geq 2$.

Proof. We assume that the architecture consists of only one recursive layer; maybe we have to introduce a time delay as constructed in Lemma 4.3.1. Assume that s inputs of height at most t are shattered. Then the different transition functions transforming an input and context to a new context, which each correspond to a recursive network, act on a domain of size $D = 2^{Nk} \cdot sk^t$ (= different activations of the context neurons $\cdot$ maximum number of different inputs). Note that the latter term is not affected by a time delay. Each neuron with W_i weights implements at most $2D^{W_i}$ different mappings on the input domain [132] (Theorem 4.1, Example 4.3), consequently, the entire hidden layer implements at most $2^N D^W$ different transition functions. Now s inputs are shattered, therefore $2^s \leq 2^N (2^{Nk} sk^t)^W$, which leads to the bound $s = O(NWk + tW \ln k + W \ln W)$. $\qquad\square$

For large t this is a better bound than the previous one since the VC-dimension only grows linearly with increasing height. In particular, the same argumentation shows that for inputs from a finite alphabet Σ the VC-dimension is finite, too, since in this case the term sk^t, which describes the different inputs, reduces to the term $|\Sigma|^{Nk}$.

Corollary 4.3.1. *If the size of the input alphabet Σ is finite, the bound $\mathcal{VC}(\mathcal{F}) = O(NkW(\ln|\Sigma| + 1))$ is valid for $\sigma = \text{H}$.*

But unfortunately, the bounds depend in most interesting cases on the maximum input height t and therefore become infinite if the input set consists of trees with unbounded height. The lower bounds on the VC-dimension obtained by Koiran and Sontag [70] indeed show that the dependence of these bounds on t is necessary, and the order of the dependence corresponding to t is tight in the case $k = 1$, $\sigma = H$. But since Koiran and Sontag restrict their considerations to networks with only one input neuron the lower bounds can be slightly improved for general networks.

Theorem 4.3.2. *If $\sigma = H$ and $k = 1$ then $VC(\mathcal{F}|X_t) = \Omega(W \ln(Wt))$.*

Proof. In [70] a recurrent architecture shattering $\Omega(W \ln(t/W))$ points is constructed, in [82] the existence of a feed-forward architecture shattering $\Omega(W \ln W)$ points is shown. Any feed-forward architecture can be seen as a special recurrent network where the context units have no influence on the computation, and the corresponding weights are 0. We can take one architecture with W weights shattering $\Omega(W \ln(t/W))$ points, two architectures each with W weights and shattering $\Omega(W \ln W)$ points and combine them in a single architecture with $3W + 28$ weights which shatters $\Omega(W \ln(tW))$ points as desired. The technical details of the combination are described in the next lemma. $\qquad\square$

Lemma 4.3.5. *For feed-forward architectures $f^1 : \mathbb{R}^{n_1+k} \to \{0,1\}$, ..., $f^l : \mathbb{R}^{n_l+k} \to \{0,1\}$ with $w_1, \ldots, w_l$ weights such that the corresponding induced architectures f_0^i shatter t_i input trees, a folding architecture with $w_1 + \ldots + w_l + 9l + 1$ weights can be constructed shattering $t_1 + \ldots + t_l$ input trees of the same height as before. The architecture consists of a combination of the single architectures with some additional neurons which are equipped with the perceptron activation function.*

Proof. Here we denote the architectures by the symbols of a typical function which they compute. Define $f : \mathbb{R}^{n_1+\ldots+n_l+1+kl} \to \{0,1\}^l$,

$$f(x_1, \ldots, x_l, x, y_1, \ldots, y_l) = \\ (f^1(x_1, y_1) \wedge x = b_1, \ldots, f^l(x_l, y_l) \wedge x = b_l)$$

for pairwise different $b_i \in \mathbb{R}$. f can be implemented with the previous $w_1 + \ldots + w_l$ weights and $8l$ additional weights in $3l$ additional perceptron units. Now if we expand each label a in a tree t of the t_i trees that are shattered by $\tilde{f}_0^i$ to a vector $(\;\underbrace{0, \ldots, 0}_{n_1+\ldots+n_{i-1}}\;, a, \underbrace{0, \ldots, 0}_{n_{i+1}+\ldots+n_l}\;, b_i)$, the induced architecture $\tilde{f}_0$ computes the output $(0, \ldots, \tilde{f}_0^i(t), \ldots, 0)$. Consequently, the folding architecture which combines $\tilde{f}_0$ with an OR-connection shatters the $t_1 + \ldots + t_l$ input trees obtained by expanding the inputs shattered by the single architectures $f^1, \ldots, f^l$ in the above way. The height of the trees remains the same, and this architecture has $w_1 + \ldots + w_l + 9l + 1$ weights. See Fig. 4.3. $\qquad\square$

Obviously, the lower bounds for $k = 1$ hold for the cases $k \geq 2$, too, but except for a linear or polynomial activation function, the upper bounds differ from these lower bounds by an exponential term in t. We can improve the lower bound for the perceptron activation function as follows:

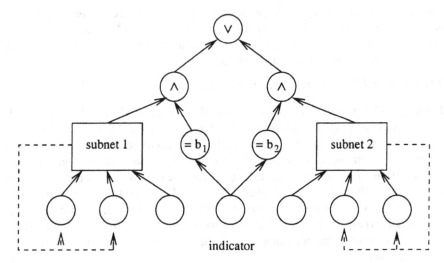

Fig. 4.3. Combining architectures to a single architecture with corresponding VC-dimension

Theorem 4.3.3. *If $\sigma = H$ and $k \geq 2$ then $VC(\mathcal{F}|X_t) = \Omega(Wt + W \ln W)$.*

Proof. We assume $k = 2$, the generalization to $k > 2$ follows directly. We write the 2^{t-1} dichotomies of $t - 1$ numbers as lines in a matrix and denote the ith column by e_i. e_i gives rise to a tree of height t as follows: In the 2^{t-1} leaves we write the components of $e_i + (2, 4, 6, \ldots, 2^t)$; the other labels are filled with 0. The function $g : \{0, 1\}^{1+2} \rightarrow \{0, 1\}$, $g(x_1, x_2, x_3) = (x_1 \in [2j + 1/2, 2j + 3/2]) \vee x_2 \vee x_3$ can be computed by a feed-forward network with activation function H with 11 weights including biases. g induces a function $\tilde{g}_0$ which computes just the component j of e_i for the input tree corresponding to e_i. Since any dichotomy of the $t - 1$ trees can be found as a line of the matrix with columns e_i the trees corresponding to the vectors e_i can be shattered by the corresponding architecture.

For any $w \in \mathbb{N}$ a feed-forward architecture h exists with activation function H and w weights which shatter $\Omega(w \ln w)$ points [82]. We can trivially expand the input dimension of this architecture by 2 such that – denoting this architecture by the same symbol – the induced mapping $\tilde{h}_0$ coincides with the original mapping h on input trees of height one.

We want to combine several copies of $\tilde{g}_0$ and $\tilde{h}_0$ such that all sets shattered by each single architecture are shattered by only one new architecture. Using Lemma 4.3.5 we obtain a folding architecture which shatters $\Omega(Wt + W \ln W)$ trees of height at most t and which has no more than W parameters if we combine $\lfloor (W - 20)/40 \rfloor$ architectures induced by g each shattering $t - 1$ trees of height t and one architecture induced by h with $\lfloor W/2 \rfloor$ weights shattering $\Omega(W \ln W)$ points. Note that the input dimension could be reduced because the shattered inputs have labels equal to 0 at most places. In fact, the architectures corresponding to $\tilde{g}_0$ could share their inputs.   □

Note that the order of this bound is tight for $t > W$. In the sigmoidal case we are unfortunately only aware of a slight improvement of the lower bound such that it still differs from the upper bound by an exponential term.

Theorem 4.3.4. *For $k \geq 2$ and $\sigma = $ sgd an architecture exists shattering $\Omega(t^2 W)$ points.*

Proof. As before we restrict the argumentation to the case $k = 2$. Consider the $t(t+1)/2$ trees of depth $t + 1$ in $(\mathbb{R}^2)_2^*$ which contain all binary numbers $\underbrace{0 \ldots 0}_{t}$ to $\underbrace{1 \ldots 1}_{t}$ of length t in the first component of the labels of the leaves, all binary numbers of length $t - 1$ in the labels of the next layer, ..., the numbers 0 and 1 in the first layer, and the number 0 in the root. In the tree t_{ij} ($i \in \{1, \ldots, t\}$, $j \in \{1, \ldots, i\}$) the second component of the labels is 0 for all except one layer $i + 1$ where it is 1 at all labels where the already defined coefficient has a 1 as the jth digit. $t_{2,1}$ is the tree $(0,0)((0,0)((00,0),(01,0)),(1,0)((10,1),(11,1)))$ if the depth $t + 1$ is 3, for example.

The purpose of this definition is that the coefficients which enumerate all binary strings are used to extract the bits number $1, \ldots, t(t+1)/2$ in an efficient way from the context vector: We can simply compare the context with these numbers. If the first bits correspond, we cut this prefix by subtracting the number from the context and obtain the next bits for the next iteration step. The other coefficient of the labels specify the digit of the context vector which is responsible for the input tree t_{ij}, namely the $1 + \ldots + i - 1 + j$th digit. With these definitions a recursive architecture can be constructed which just outputs for an input t_{ij} the responsible bit of the initial context and therefore shatters these trees by an appropriate choice of the initial context.

To be more precise, the architecture is induced by the mapping $f : \mathbb{R}^{2+6} \rightarrow \mathbb{R}^3$,

$$f(x_1, x_2, y_1, y_2, y_3, z_1, z_2, z_3) =$$
$$(\max\{1_{y_1 y_2 - x_1 \in [0,1[} \cdot (y_1 y_2 - x_1), 1_{z_1 z_2 - x_1 \in [0,1[} \cdot (z_1 z_2 - x_1)\},$$
$$0.1 \cdot y_2, y_3 \vee z_3 \vee (x_2 \wedge y_1 y_2 - x_1 \in [0,1[) \vee (x_2 \wedge z_1 z_2 - x_1 \in [0,1[)),$$

which leads to a mapping which computes in the third component the responsible bit of y for t_{ij} with an initial context $(y = 0.y_1 y_2 \ldots y_{t(t-1)/2}, (10)^t, 0)$. The role of the first context neuron is to store the remaining bits of the initial context, at each recursive computation step the context is shifted by multiplying it by y_2 and dropping the first bits by subtracting an appropriate label of the tree in the corresponding layer. The second context neuron computes the value $10^{\text{height of the remaining tree}}$. Of course, we can substitute this value by a scaled version which is contained in the range of sgd. The third context neuron stores the bit responsible for t_{ij}: To obtain an output 1 the first bits

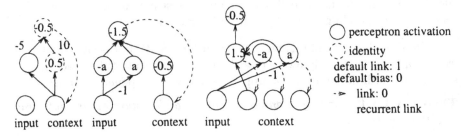

Fig. 4.4. Networks with infinite VC-dimension. Left: Sigmoidal network. Middle: Perceptron network. Right: Recurrent cascade correlation network

of an appropriate context have to coincide with a binary number which has an entry 1 at the position that is responsible for the tree. This position is indicated by x_2. f can be approximated arbitrarily well by an architecture with the sigmoidal activation function with a fixed number of neurons. It shatters $t(t+1)/2$ trees.

By first simulating the initial context with additional weights, as described in Lemma 4.3.3, and adding W of these architectures, as described in Lemma 4.3.5 where we approximate the perceptron activation by the sigmoidal function, we obtain a folding architecture shattering $\Omega(t^2 W)$ trees.   □

The same argumentation holds for any activation function which can approximate the perceptron activation function, the identity, and square activation function; for example, a piecewise polynomial function with two constant pieces and at least one quadratic piece.

Some aspects occurring in the constructions of the other lower bounds for recurrent networks are worth mentioning: The network depicted in Fig. 4.4 on the left with starting vector $0.y_1 \ldots y_n 1 - 0.5$, $y_i \in \{0, 5\}$ computes the output $0.y_i \ldots y_n 1 - 0.5$ in the ith recurrent step. Therefore the corresponding sigmoidal recurrent architecture with only 3 computation neurons, which approximates the perceptron activation by $\mathrm{sgd}(x/\epsilon)$ ($\epsilon \to \infty$), and the identity by $(\mathrm{sgd}(\epsilon x) - 0.5)/(\epsilon \cdot \mathrm{sgd}'(0))$ ($\epsilon \to 0$), shatters any input set where each input has a different height. This input set occurs if the input patterns are taken from one time series so that the context length increases for any new pattern. That is, in the sigmoidal case very small architectures shatter inputs which occur in practical learning tasks.

In the case of the perceptron activation function the architectures with infinite VC-dimension are also very small, but the set that is shattered stores the possible dichotomies in the sequence entries, that is, it has a special form and does not necessarily occur in practical learning tasks. The same is valid if we argue with the dual VC-dimension. But nevertheless, an argumentation with the dual VC-dimension can shed some light on other aspects: The class

$$\mathcal{F} = \{f_a : \mathbb{R}^* \to \{0,1\} \mid f_a(x) = 1 \iff a \text{ occurs in the sequence } x, a \in \mathbb{R}\}$$

can be implemented by a recurrent perceptron architecture with 4 computation neurons, as depicted in Fig. 4.4 in the center. We can consider $\mathcal{F}$ as a function class which is parameterized by a. $\mathcal{F}^\vee$ restricted to inputs of length t has VC-dimension $\Omega(t)$. Consequently, recurrent networks with 4 computation neurons and perceptron activation function have VC-dimension $\Omega(\ln t)$.

Note that we need only one self-recurrence, therefore the same construction works for a recurrent cascade architecture [33]. In a recurrent cascade network the feed-forward function h may depend on the inputs, too. The recurrent part does not contain hidden neurons. The function computing the state $s_i(t)$ at time t of the ith context neuron has the form $\sigma(g_i(s_i(t-1), s_{i-1}(t), \ldots, s_1(t), \text{input}))$ with a linear function g_i and an activation function σ (see Fig. 4.4 on the right).

It is quite natural to assume that a function class used for time series prediction, where the last few values are not the important ones and a moving window technique and standard feed-forward networks are not applicable, can at least tell whether a special input has occurred or not. As a consequence any appropriate function class in this learning scenario contains the class $\mathcal{F}$ we defined above as a subclass – and has therefore infinite VC-dimension. In other words, even very restricted function classes lead to an infinite VC-dimension due to the recursive structure of the inputs.

4.3.3 Lower Bounds on $fat_\epsilon(\mathcal{F})$

One solution of this dilemma can be to restrict the number of different possible events to a finite number which leads to a finite VC-dimension in the perceptron case, as already mentioned. In the sigmoidal case we can try another escape from the dilemma: As stated earlier, the interesting quantity in learning tasks concerning real valued functions is the fat shattering dimension and not the pseudo-dimension. The fat shattering dimension is limited by the pseudo-dimension, therefore the upper bounds transfer to this case. But function classes exist with infinite pseudo-dimension and finite fat shattering dimension which are efficiently used as a learning tool, for example, in boosting techniques [44]. For arbitrary weights the fat shattering dimension of networks coincides in fact with the pseudo-dimension – we can simply scale the output weights to obtain an arbitrary classification margin. We therefore restrict the weights. In a learning procedure this is implemented using weight decay, for example. At least in the linear case it is necessary to restrict the inputs, too, since a division of the weights by a real number x corresponds to a scaling of the input at time i by x^{t-i+1} if the maximum length is t and the initial context is 0 in recurrent networks.

Of course, in the perceptron case a restriction of the weights and inputs has no effect since the test as to whether a unit activation is ≥ 0 is computed with arbitrary precision. This assumption may be unrealistic and is dropped if we approximate the perceptron function by a continuous one.

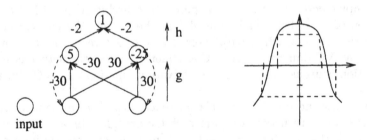

Fig. 4.5. Left: sigmoidal network with infinite fat shattering dimension. Right: function leading to an infinite fat shattering dimension if applied recursively to an appropriate initial context.

The constructions leading to the lower bounds use an approximation process in the sigmoidal case resulting in unlimited weights, and they use arbitrarily growing inputs in the linear case [27, 70]. Now we assume that the inputs are taken from $[-1, 1]$ and the weights are absolutely bounded by B. We fix the parameter for the fat shattering as $\epsilon > 0$. Unfortunately one can find other architectures with limited weights and inputs, but infinite fat shattering dimension for unlimited input length t.

Theorem 4.3.5. *The ϵ-fat shattering dimension for $\epsilon \leq 0.4$ of sigmoidal recurrent networks with three computation neurons, inputs in $[-1, 1]$, and weights absolutely bounded by $B \geq 30$, is at least linear in the length of the input sequences.*

Proof. The mapping $f(x) = 1 - 2 \cdot \text{sgd}(15x - 10) - 2 \cdot \text{sgd}(-15x - 10)$ maps the interval $[-0.9, -0.4]$ to an interval containing $[-0.9, 0.9]$, and in the same way $[0.4, 0.9]$. Therefore for any sequence $\epsilon_1, \ldots, \epsilon_n$ of signs we can find a real value $y \in [-1, 1]$ so that $f^i(y) \in \begin{cases} [-0.9, -0.4] & \text{if } \epsilon_i = + \\ [0.4, 0.9] & \text{if } \epsilon_i = - \end{cases}$. We start with a value according to the sign of ϵ_n. Afterwards, we recursively choose one of the at least two inverse images of this value. One sign of the inverse images corresponds to the preceding ϵ_i. The iterated mapping f can be simulated by a recurrent network with 2 context neurons, as shown in Fig. 4.5. All weights and inputs are bounded and any set of sequences of different height can be shattered by the corresponding architecture since we can choose the signs ϵ_i in the procedure to find the starting vector y according to the dichotomy. In the network in Fig. 4.5 the initial context is changed to, e.g., $((1 - y)/2, 0)$ instead of y which is still bounded by B. $\square$

For W weights a lower bound $\Omega(tW)$ can be derived by composing a constant fraction of W of such architectures, as described in Lemma 4.3.5. Since the equality which is used in the proof of 4.3.5 is to be approximated with a sigmoidal network, only a finite number of networks can be combined in such a way without increasing the weights.

Since only the length of the input sequences is important the shattered set occurs in practical learning tasks as before. Furthermore, the same argumentation is valid for any function where a linear combination leads to a similar graph as in Fig. 4.5, with the property that a positive and a negative interval P and N exist with $N \cup P \subset f(N) \cap f(P)$. In particular, this argumentation holds with possibly different weight restrictions for any activation function which is locally C^2 with a nonvanishing second derivative. It seems that the possibility of chaotic behavior is responsible for the infiniteness of the fat shattering dimension since, regardless of the special input, any sequence of signs can be produced by an appropriate choice of the initial context.

But even in the case of linear activation the fat shattering dimension remains infinite. We argue with the dual class, consequently, the shattered set is more special in this case and stores in some sense the dichotomies we want to implement.

Theorem 4.3.6. *The ϵ-fat shattering dimension for $\epsilon \leq 0.3$ of recurrent architectures with linear activation and 2 computation units is $\Omega(\ln^2 t)$ if t denotes the maximum input length, the weights are absolutely bounded by 2, and the inputs are contained in $[-1, 1]$.*

Proof. For an angle $\varphi = \pi/n$ the sequence $\sin(i\varphi + \pi/(2n))$, $i = 0, 1, \ldots$ is the solution of the difference equation

$$x_0 = \sin\left(\frac{\pi}{2n}\right), \ x_1 = \sin\left(\frac{3\pi}{2n}\right), \ x_{i+2} = 2\cos(\varphi)x_{i+1} - x_i.$$

The difference equation can be implemented by a recurrent neural architecture with bounded weights and 2 computation units, that is, the output for an arbitrary input sequence of length i is $\sin(i\varphi + \pi/(2n))$. But then we can produce for any natural number n a sequence of n consecutive positive values, n consecutive negative values, and so on. We fix t networks which produce these alternating signs for $n = 1, 2, 4, 8, \ldots, 2^{t-1}$. Any dichotomy of these t networks appears as an output vector of the t networks at one of the time steps $1, \ldots, 2^t$. Therefore these networks can be shattered if one chooses an input of accurate length – the dual class of the linear recurrent architecture has VC-dimension of at least t for inputs of length 2^t even if the weights and inputs are bounded.

We are interested in the fat shattering dimension. For growing n in the previous construction the signs are correct but the value $\sin(i\pi/n + \pi/(2n))$ becomes small if i equals multiples of n. Therefore we have to modify the construction. We simply consider longer input sequences and ignore the outputs that are too small: We requested |output| $\geq \epsilon \leq 0.3$. Since $0.3 \leq \sin(\pi/8 - 1/16)$ for a sign sequence of $n = 2^i$ consecutive equal signs for $i \geq 3$ the first $2^i/8$ values are too small, and even so the same number before and after any sign change. For an even number t we consider the $3t/2$ networks producing alternating signs for $n = 1, 2, \ldots, 2^{3t/2-1}$ as above. We

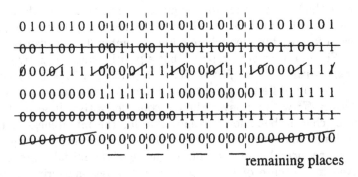

Fig. 4.6. Example of output canceling: The time steps where an output of a network is too small are canceled in the consideration.

drop the networks number $2, 5, 8, \ldots, 3t/2 - 1$ and in the remaining networks the outputs corresponding to the time steps $1, \ldots, 2^{3i}/8$ before and after any sign change in the network number $3i$ for $i = 1, \ldots, t/2$. This means that in any three steps we cancel half the number of outputs balanced around the sign changes in the actual network. Fig. 4.6 shows an example.

In the networks number $1, 4, 7, \ldots$ we revert the signs. Still any dichotomy of the networks is stored in the networks' outputs at one time step in the remaining t networks and output sequences with original length $2^{3t/2}$ with 2^t single outputs that are not canceled. These outputs have an absolute value of at least ϵ. This follows by induction on $i \leq t$:

In the network $3i$ there are 2^{2i} of any 2^{3i} outputs not canceled since in the $3(i + 1)$st step of 8 consecutive runs of 2^{3i} outputs with equal sign, the first, fourth, fifth, and last run is dropped, that is, there are $8 \cdot 2^{2i} - 4 \cdot 2^{2i} = 2^{2(i+1)}$ outputs left. In the $3i$th step all sign sequences with $n \leq 2^{2i-1}$ can be found since from step $3i$ to $3(i + 1)$ the old sign sequences are not disturbed, and additionally, the sequence with 7 sign changes in $2^{3(i+1)}$ original outputs leads to the reverted sequence with $n = 2^{2i}$, the sequence with one sign change leads to the sequence with $n = 2^{2i+1}$. The absolute values of the outputs are correct since if we drop the values $1, \ldots, n/4$ of n consecutive signs, $n/8 \leq n/4$ and the same is valid for $2 \cdot n$ instead of n on the left side.

We have found t networks where any dichotomy of the networks correspond to one output at time $\leq 2^{3t/2}$ of the networks. Therefore the dual class, that is, the recurrent architecture with inputs of length at most t has ϵ-fat shattering dimension $\Omega(\ln^2 t)$. The additional log-term is due to the fact that we considered the dual VC dimension. $\qquad \square$

We can combine W of such architectures to obtain a lower bound $\Omega(W \ln^2 t)$ as follows: We simulate the initial context by two additional weights and input neurons which receive the input value 1 at the first step and 0 at all other steps (compare Lemma 4.3.3). We take W of these architectures and combine the single output neurons with one linear output neuron. If we consider inputs where the values which simulate the initial contexts are 0 at any time step for all but one architecture, then only the output of this architecture differs from 0. Consequently, all single architectures can be simulated within this combination by an appropriate choice of the additional inputs.

Here the reasons for the infiniteness are, generally, that even for very uniform outputs as $-1^n 1^n \ldots$, for example, one can find points where any output value is possible in an appropriate sequence. The information that can be stored in the input sequences such that a network can use it in an easy way is not bounded if the length of the sequences is not bounded. As a consequence even the fat shattering analysis leads to lower bounds which depend on the input length in all interesting cases.

4.4 Consequences for Learnability

It follows from the bounds on the various dimensions that distribution-independent learnability cannot be guaranteed for folding architectures under realistic conditions. The natural hierarchy of the class described by the number of parameters collapses if the corresponding capacity is considered. As a consequence it is necessary to find other guarantees for the generalization capability. The learning process and structural risk minimization has to take some different stratification into account.

When dealing with distribution-dependent learnability, bounds on the number of examples which guarantee valid generalization can be found. This is due to the fact that a natural stratification of the learning scenario is given via the input set. Since the capacity of folding architectures with restricted inputs is bounded in dependence on the number of parameters and the input length, we can apply Theorem 4.2.8 to this situation and obtain bounds which depend on the special distribution.

Corollary 4.4.1. *Assume p_t is the probability of inputs of height at most t. Then any learning algorithm for a folding architecture which produces small empirical error is PAC. The number of examples necessary for valid generalization for the minimum risk algorithm is polynomial in the desired accuracy if p_t is of the order $1 - d_t^{-\beta}$ for some $\beta > 0$. d_t is the VC- or pseudo-dimension of the folding architecture restricted to inputs of height at most t.*

As a consequence we obtain exponential bounds if the probability of high trees vanishes with increasing height in a less than logarithmical manner in the case of $\sigma = $ H, $k = 1$ or $\sigma = $ id and $k \geq 1$, or in a less than polynomial manner in the case of $\sigma = $ H, $k \geq 2$ or σ a polynomial, $k \geq 1$ or $\sigma = $ sgd, $k = 1$, or in a less than exponential manner in the case $\sigma = $ sgd, $k \geq 2$.

But this observation does not necessarily lead to the consequence that any concrete example exists where the training set increases exponentially.

Indeed, the existence of a uniform bound may be impossible for a class of distributions, although each single distribution guarantees polynomial learnability, as can be seen in the following example:

Example 4.4.1. There exists a class $\mathcal{D}$ of probability distributions and a function class $\mathcal{F}$ such that for any learning algorithm no uniform upper bound $m = m(\epsilon, \delta)$ exists with

$$\inf_{P \in \mathcal{D}} \sup_{f \in \mathcal{F}} P^m(\mathbf{x} \mid d_P(f, h_m(f, \mathbf{x})) > \epsilon) \leq \delta,$$

but for any single distribution $P \in \mathcal{D}$ PAC learnability with a polynomial number of examples is guaranteed.

Proof. Define $F_t = \{f_1, \ldots, f_t\}$ with

$$f_l : [0, 1] \to \{0, 1\}, \quad f_l(x) = \begin{cases} 0 & \text{if } x \in \displaystyle\bigcup_{i=0}^{2^{l-1}-1} \left[\frac{2i}{2^l}, \frac{2i+1}{2^l}\right] \\ 1 & \text{otherwise.} \end{cases}$$

These are almost the same functions as in Example 4.2.6, except that a perceptron activation function is added at the outputs. For the uniform distribution P_u on $[0, 1]$ it holds that $d_{P_u}(f_i, f_j) = 0.5$ for $i \neq j$. For $x \in [0, 1]$ define a tree $t_t(x)$ of height t as follows: The ith leaf ($i = 1, \ldots, 2^{t-1}$) contains the label $(x/(2i - 1), x/(2i)) \in \mathbb{R}^2$, any other label is $(0, 0)$. Any function $f_l \in F_t$ can be computed as the composition of the injective mapping t_t with $\tilde{g}_0$ where $g : \mathbb{R}^2 \times \{0, 1\}^2 \to \{0, 1\}$, $g(x_1, x_2, x_3, x_4) = x_3 \vee x_4 \vee (x_1 > 1/2^l \wedge x_2 < 1/2^l)$ and l is chosen according to the index of f_l. Let $\mathcal{F}$ denote the folding architecture corresponding to g. Then $M(\epsilon, \mathcal{F}|X'_t, d_{P_t}) \geq t$ for $\epsilon < 0.5$, if X'_t denotes the set of trees of exactly height t and P_t denotes the probability on X'_t induced by t_t and the uniform distribution P_u on $[0, 1]$. Now $\mathcal{D}$ contains all distributions D_t, where D_t equals P_t on X'_t and assigns 0 to all other trees.

Since $M(\epsilon, \mathcal{F}, dD_t) \geq t$, at least $\lg(t(1 - \delta))$ examples are needed to learn the functions correctly with respect to D_t and confidence δ. Obviously, this number exceeds any uniform bound for m. On the contrary, any single distribution D_t ensures PAC learnability with a polynomial number of examples of the order $d/\epsilon \ln(1/\epsilon)$, where $d = \mathcal{VC}(\mathcal{F}|X_t)$ for any consistent learning algorithm. $\square$

Since the bounds which are obtained for each single distribution D_t in the above example are not uniform with respect to t the above example is not really surprising. A very sophisticated example for a class of probability distributions such that each single distribution allows learnability with uniform bounds but the entire class of probabilities is not learnable can be found in [132] (Example 8.1). However, we can use the above construction to obtain a witness for another situation: It is of special interest whether one single distribution can be found where the sample size necessarily grows exponentially. Note that in distribution-independent learning such a situation cannot exist since concept classes are either distribution-independent learnable and have finite VC-dimension, which implies polynomial bounds, or they are not distribution-independent learnable at all. The following construction is an example of an exponential scenario. This in particular answers a question which Vidyasagar posed in Problem 12.6 [132].

Example 4.4.2. A concept class exists where the number of examples necessary for valid generalization with accuracy ϵ is growing exponentially with $1/\epsilon$.

Proof. Using Lemma 4.1.1 it is sufficient to show that the covering number of the function class to be learned grows more than exponentially in $1/\epsilon$. The idea of the construction is as follows: The input space X is a direct sum of subspaces, *i.e.*, the trees of height t. On these subspaces we use the construction from the previous example which corresponds to function sets with a growing covering number. But then we can shift the probabilities of the single subspaces so that trees with rapidly increasing heights have to be considered to ensure a certain accuracy.

Define $\mathcal{F}$, X'_t, and P_t as in the previous example. Since X is the disjoint union of X'_t for $t = 1, \ldots, \infty$, we can define a probability P on X by setting $P|X'_t = P_t$ and choosing $P(X'_t) = p_t$ with

$$p_t = \begin{cases} 6/(n^2\pi^2) & \text{if } t = 2^{2^n} \text{ for } n \geq 1, \\ 0 & \text{otherwise.} \end{cases}$$

Since $d_P(f, g) \geq p_t d_{P_t}(f|X'_t, g|X'_t)$ it is

$$M(\epsilon, \mathcal{F}, d_P) \geq M(\sqrt{\epsilon}, \mathcal{F}|X'_t, d_{P_t}) \geq t$$

for $1/2 > \sqrt{\epsilon}$ and $p_t \geq \sqrt{\epsilon}$. Assume $\epsilon = 1/n$. Then $p_t \geq \sqrt{\epsilon}$ is valid for $t = 2^{2^m}$, $m \leq \sqrt{6}/\pi\, n^{1/4}$. As a consequence

$$M(1/n, \mathcal{F}, d_P) \geq 2^{2^{c \cdot n^{1/4}}}$$

for a positive constant c. This number is growing more than exponentially in n. $\qquad\qquad\square$

Unfortunately, the argumentation of Corollary 4.4.1 needs prior information about the underlying probability. Only if the probability of high trees is restricted and therefore the recurrence is limited can *a priori* correct generalization be guaranteed. In contrast, the luckiness framework allows us to train situations with arbitrary probabilities, estimate the probability of high trees a posteriori from the training data, and output a certain confidence of the learning result which depends on the concrete data.

Corollary 4.4.2. *Assume $\mathcal{F}$ is a $[0,1]$-valued function class on the trees, P is a probability distribution, $d_t = \mathcal{PS}(\mathcal{F}|$ trees of height $\leq t)$, and p_i are positive numbers with $\sum_i p_i = 1$; then the unluckiness function $L'(\mathbf{x}, f) = \max\{$height of a tree in $\mathbf{x}\}$ leads to a bound*

$$\sup_{f \in \mathcal{F}} P(\mathbf{x} \mid |\hat{d}_m(f, h_m(f, \mathbf{x}), \mathbf{x}) - d_P(f, h_m(f, \mathbf{x}))| \leq \epsilon) \geq 1 - \delta$$

for

$$\epsilon = 4\alpha + \frac{8\lg(1/\delta)}{m} + \sqrt{\frac{16}{m}\left(2\lg\left(\frac{1}{\delta}\right)\ln m + (3+i)\ln 2 + \ln\frac{4}{p_i\delta}\right)}$$

where $i \geq d_{L'(\mathbf{x}, h_m(f,\mathbf{x}))} \lg (4em/\alpha \cdot \ln(4em/\alpha))$ and $\alpha > 0$.

Proof. L' is smooth with respect to $\Phi(m, h, \delta, \alpha) = 2\left(4em/\alpha \cdot \ln(4em/\alpha)\right)^d$, where $d = \mathcal{PS}(\mathcal{F} \mid$ trees of height $\leq L'(\mathbf{x}, h_m(f, \mathbf{x})))$ and $\eta(m, h, \delta, \alpha) = \ln(1/\delta)/(m\ln 2)$, as can be seen as follows:

$$P^{2m}(\mathbf{xy} \mid \exists f \in \mathcal{F} \, \forall \mathbf{x'y'} \subset_\eta \mathbf{xy} \; l(\mathbf{x'y'}, f, \alpha) > \Phi(m, L'(\mathbf{x}, h), \delta, \alpha))$$
$$\leq \quad P^{2m}(\mathbf{xy} \mid \min_{\mathbf{x'y'} \subset_\eta \mathbf{xy}} \max\{\text{height in } \mathbf{x'y'}\} > \max\{\text{height in } \mathbf{x}\})$$

because the number of functions in $|\{f|\mathbf{x'y'} \mid f \in \mathcal{F}\}_\alpha|$, where $\mathbf{x'y'}$'s height is at most $L'(\mathbf{x}, h_m(f, \mathbf{x}))$, can be bounded by $\Phi(m, L'(\mathbf{x}, h), \delta, \alpha)$ because the number is bounded by $M(\alpha/(2m), (\mathcal{F}|\mathbf{x'y'})_\alpha, \hat{d}_{m'})$ where $m' = $ length of $\mathbf{x'y'}$. The latter probability equals

$$\int_{X^{2m}} U(\sigma \mid (\mathbf{xy})^\sigma \in A) dP^{2m}(\mathbf{xy}),$$

where U is the uniform distribution on the swapping permutations of $2m$ elements and A is the above event. We want to bound the number of swappings of $\mathbf{xy}$ such that on the first half no tree is higher than a fixed value t, whereas on the second half at least $m\eta$ trees are higher than t. We may swap at most all but $m\eta$ indices arbitrarily. Obviously, the above probability can be bounded by $2^{-m\eta}$, which is at most δ for $\eta \geq \lg(1/\delta)/m$.

Furthermore, the condition $\Phi(m, L(\mathbf{x}, h_m(f, \mathbf{x})), \delta, \alpha) \leq 2^{i+1}$ leads to a bound for i. Now we can insert these values into the inequalities obtained at the luckiness framework and get the desired bound. $\qquad\square$

Obviously, the bound for ϵ tends to 4α for large m. But we still need a good prior estimate of the probabilities p_i for good bounds. Furthermore, we have not allowed dropping a fraction in $\mathbf{x}$, either, when measuring the maximum height of the example trees. If the sample size increases this would be useful since the fraction of large trees is described by the probability of those trees. This probability may be small but not exactly 0 in realistic cases. Unfortunately, allowing to drop a fraction η_h in $\mathbf{x}$ when measuring L' would lead to a lower bound η_h for η. Then the factor $\eta \ln m$ in ϵ would tend to ∞ for increasing m.

4.5 Lower Bounds for the LRAAM

The VC analysis of recurrent networks has an interesting consequence for the LRAAM. It allows us to derive lower bounds on the number of neurons that are necessary for appropriate decoding with the dynamics, as introduced in section 2.3. Assume a concrete architecture is capable of appropriately encoding and decoding trees of height t such that the composition yields the identity of those trees. We restrict ourselves to symbolic data, *i.e.*, the labels are contained in the alphabet $\{0, 1\}$. Instead of exact decoding we only require approximate decoding. Hence labels > 0.5 are identified with 1, labels < 0.5 are identified with 0. Then the possibility of proper encoding yields the following result:

Theorem 4.5.1. *Assume points in $\mathbb{R}^m$ exist which are approximately decoded to all binary trees of height at most t with labels in $\{0, 1\}$ with some $\bar{h}_Y$, $h = (h_0, h_1, h_2) : \mathbb{R}^m \to \mathbb{R}^{1+m+m}$, $Y \subset \mathbb{R}^m$. If h is a feed-forward neural network, then the number of neurons is lower bounded by $2^{\Omega(T)}$ if the activation function is the standard sigmoidal activation function or a piecewise polynomial activation.*

Proof. $h = (h_0, h_1, h_2) : \mathbb{R}^m \to \mathbb{R} \times \mathbb{R}^m \times \mathbb{R}^m$ gives rise to a recurrent network $\tilde{g}_y : \mathbb{R}^* \to \mathbb{R} \times \mathbb{R}^m$, $g : \mathbb{R} \times \mathbb{R} \times \mathbb{R}^m \to \mathbb{R} \times \mathbb{R}^m$,

$$(x_0, x_1, x_2) \mapsto (h_0(x_1), (1 - x_0) \cdot h_1(x_2) + x_0 \cdot h_2(x_2)).$$

If $\bar{h}_Y$ maps value z to some tree t, then $\pi_1 \circ \tilde{g}_{(0, z)}$ maps any binary sequence of length i to some node in the ith level of the tree t, π_1 being the projection to the first component; the exact number of the node depends on the sequence: $[0, \ldots, 0]$ is mapped to the leftmost node in the ith level, $[1, \ldots, 1]$ is mapped to the rightmost node, the other sequences lead to the nodes in between. The last component of the sequences is not relevant; see Fig. 4.7 .

If points in $\mathbb{R}^m$ exist that are approximately mapped to all trees of height t in $\{0, 1\}_2^*$ with $\bar{h}_Y$, the neural architecture $\pi_1 \circ \tilde{g}_{(0, \_)}$ shatters all binary sequences of length t with the last component 1: One can simply choose the second part of the initial context corresponding to a vector z which encodes a tree of height t and leaves according to the dichotomy.

g can be computed by adding a constant number of neurons with some at most quadratic activation function to h and it can be approximated arbitrarily well adding a constant number of sigmoidal units to h. Consequently, the VC-dimension of $\tilde{g}_{(0, \_)}$, restricted to inputs of height at most t, is limited by $O(N^3 T \ln(qd))$ if the activation function in h is piecewise polynomial with at most q pieces and degree at most $d \geq 2$. The VC-dimension is limited by $O(N^4 T^2)$ if the activation function is the standard sigmoidal function. In both cases, N denotes the number of neurons in h. The lower bound 2^{T-1} for the VC-dimension leads to bound $N = 2^{\Omega(T)}$ for the neurons in h. $\square$

Note that the way in which the trees are encoded in $\mathbb{R}^m$ is not important for

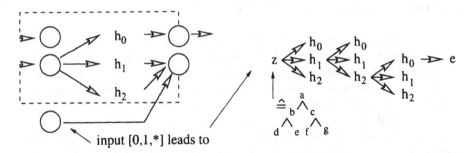

Fig. 4.7. *An appropriate input to the recurrent network as defined in Theorem 5 restores a path of the tree $h_Y(z)$, the length of the input sequence indicates the length of the path, entries 0 and 1 stand for the left or right subtree, respectively, h_0 yields the output label.*

the lower bound. Furthermore, a more sophisticated decoding of the single binary nodes or using other standard activation functions leads to the same lower bound since, for any reasonable modification, the VC-dimension of the recurrent network as defined in the proof is still bounded by some polynomial in the number of nodes and maximum input height. Consequently, the decoding formalism requires an increasing amount of resources even for purely symbolic data. Hence a formalism like the LRAAM can deal only with restricted situations, *i.e.*, almost linear trees or limited height, whereas these restrictions do not apply to methods which merely focus on encoding, like recurrent and folding networks.

4.6 Discussion and Open Questions

In this chapter we have considered the possibility of using folding networks as a learning tool from an information theoretical point of view in principle. It is necessary to investigate the question as to whether usual data sets contain enough information such that the underlying regularity can be approximated with a folding network – or it can be seen that no such regularity exists – before we can start training. If learnability is not guaranteed, no learning algorithm, however complicated, could succeed in general. Apart from guarantees of the possibility of learning in principle we have obtained concrete bounds on the number of examples sufficient for valid generalization which tell us which network size and number of patterns we should use in a concrete learning task.

In the first part of this chapter we have discussed several possibilities of formalizing PAC learnability from an information theoretical point of view. Depending on the task that only the existence of one good learning algorithm is to be proved or every consistent algorithm shall perform correct generalization, one gets the term PAC or consistent PAC learnability. If the generalization is to be uniform with respect to every function the term PAC is substituted by PUAC. Furthermore, it makes sense to consider nearly consistent algorithms with small empirical error, too, which leads to scaled versions of the above-mentioned terms. We have introduced these scale sensitive versions and examined the connection between these terms, answering in particular Problem 12.4 in [132]. One question that remains open in this context is whether a concept class exists which is PUAC learnable, but not ϵ-consistently PAC learnable for all $\epsilon > 0$.

Furthermore, it is possible to find for (almost) all concepts equivalent characterizations that do not rely on the notion of a learning algorithm. Such characterizations are of special interest for practical applications. They lead to necessary and sufficient conditions that can be tested before the design of a concrete learning algorithm. They establish the capability of any learning algorithm – maybe with some conditions – to perform correct generalization in one of the concrete forms as specified above in principle.

The well-known characterization of a finite covering number only forms a sufficient but not necessary condition for PAC learnability of function classes, due to the possibility of encoding functions uniquely in real values. Here we have shown that the argumentation no longer holds if the scaled versions of the above terms are considered because these formulations imply a certain robustness with respect to observation noise. Considering the same topic it is an interesting question as to whether the addition of noise to a learning algorithm itself causes an increase in the number of function classes that can be learned with such an algorithm. For concept classes it has been shown that all randomized algorithms can be substituted by deterministic algorithms and it can be shown that even in function learning some kind of noise, for example, a tossed coin can be simulated within a deterministic algorithm. The argument of [52], which simulates a tossed coin within a deterministic algorithm, transfers to the function case. It remains open as to whether other (reasonable) kind of noise increases the number of function classes which are PAC learnable compared to the deterministic setting.

We have briefly discussed which results remain valid if the practically interesting case of model-free learning is dealt with. This is interesting if no prior knowledge about the form of the function to be learned is available. We cited several results that deal with the distribution-independent case, where the underlying probability in accordance to which the examples are chosen is entirely unknown. Of course, there are as many open questions here as further approaches to deal with several learning scenarios that fit various kinds of practical applications.

But in concrete tasks concerning neural networks, learnability is most often established by simply examining the capacity of the function class. Finiteness of the VC- or pseudo-dimension guarantees learnability in all of the above cases.

Therefore, we have estimated the VC- or pseudo-dimension of folding architectures. Here the upper and lower bounds in the sigmoidal case unfortunately differ by an order of 2^t, where t is the maximum input height, leaving as an open problem the real order of the dimension. Furthermore, some kind of weight-sharing argument in the perceptron case would improve the bounds in the perceptron case, too. However, the VC-, pseudo-, and fat shattering dimension depend on the maximum input height even for restricted weights and inputs in most interesting cases. Furthermore, even very small architectures with only a few recurrent connections lead to such results. The input set that is shattered in the sigmoidal case, for example, occurs naturally in time series prediction.

As a consequence distribution-independent bounds for valid generalization cannot be obtained for folding architectures without further restrictions on the inputs or the function class. Furthermore, the structure of the input patterns and minimum architectures that shatter these patterns constitute situations that are likely to occur in practical learning tasks.

Taking the particular probability into account we have constructed situations where an exponential number of examples is needed for valid generalization answering Problem 12.6 in [132]. On the contrary, we have obtained natural conditions on the probability of the inputs such that polynomial bounds are guaranteed. For this purpose an approach of [3] has been generalized to function classes. But the limitation of the generalization error requires an *a priori* limitation of the probability of high trees.

Another possibility of guaranteeing valid generalization without explicit prior knowledge is to introduce a luckiness function which defines a posterior stratification of the learning scenario – here via the maximum input height in a concrete training set. For this purpose we have generalized the approach of [113] to function classes and arbitrary, not necessarily consistent, learning algorithms. Unfortunately, the bounds that are obtained in this way are rather conservative. It is not possible to substitute the maximum height of the inputs by a number such that only almost all example trees are of smaller height – which would be appropriate in our situation. From a technical point of view, this problem is due to the factor $\eta \ln m$ in the obtained bounds, which seems too conservative. Another problem with the luckiness function is the usage of the numbers p_i which measure an *a priori* confidence for an output of the algorithm with a certain luckiness. It would be natural to consider values p_i which are increasing since the confidence of obtaining an output corresponding to i_1, which is less lucky than another output corresponding to i_2, should be at least as high as the confidence of an output i_2. But for technical reasons the p_i have to add up to 1. However, we have obtained a guarantee for valid generalization in the distribution-dependent case and an a posteriori guarantee for any situation where the input trees are restricted.

In both cases the bounds should be further improved by an appropriate restriction or stratification of folding architectures. Since the reasons leading to infinite VC-dimension are twofold – the possibility of chaotic behavior in the sigmoidal case and the possibility of encoding arbitrary information in the input strings if they have unlimited length – a restriction of the architectures such that they fulfill certain stability criteria and a restriction of the inputs such that they have limited storage capacity may lead to better bounds. One simple limitation of the storage capacity is given by the accuracy of the computation, of course, which reduces any network to a finite automaton or tree automaton in a realistic computation. A more sophisticated argument for the limitation of the storage can be found if noise is taken into account. As already mentioned in the last chapter, the computational capability of folding networks reduces to tree automata if the computation is affected with noise. Of course, this leads to a finite VC-dimension, too, as carried out in [84] for recurrent networks.

Apart from the generalization capability bounds on the VC-dimension of a class which solves a specified task characterize the minimum number of resources required for a concrete implementation. This argumentation led to lower bounds for the encoding part of the LRAAM.

Chapter 5

Complexity

We have seen that the network we have chosen will be able to generalize well to unseen data; in our case the reason may be simple: Prior knowledge about the input distribution may be available telling us that high trees will rarely occur. We choose a neural architecture with an appropriate number of neurons. We start to train this architecture with one of the standard algorithms, say, back-propagation through structure. Now we wait until the algorithm has produced a network with small empirical error, hoping that the training procedure takes at most a few minutes until it succeeds. Unfortunately, the first training process gets stuck in a local minimum such that we have to start the training again. The next try suffers from a wrong choice of the training parameters and does not succeed either. The next try produces a very unstable behavior of the weight changes and stops because of a weight overflow.

Now at last, we ask whether the training algorithm we use, a gradient descent method, is appropriate for the training of folding networks. Perhaps a simple gradient descent method has numerical problems when minimizing the empirical error of a network architecture. Such a result would motivate us to introduce some heuristics or more sophisticated methods into our gradient descent or to use a minimization procedure which is based on a different idea.

Furthermore, we may investigate the complexity of training folding networks in principle. Results about the complexity of the learning task tell us whether any learning algorithm can succeed in adequate time, no matter whether it is based on a simple gradient descent method and therefore perhaps suffering from numerical problems, or whether it is based on other, more sophisticated tools. If we get results showing the NP completeness of the learning task we want to deal with, for example, we definitely know that any learning algorithm, however complicated, can take a prohibitive amount of time in some situations – unless $P = NP$, which most people do not believe.

We may ask the question about the difficulty of learning in different situations. One situation is the following setting: We fix an architecture and a training set and try to minimize the empirical error for this single architecture. The architecture may have several layers and use a certain activation function, in practical applications usually a sigmoidal function. Then of course, it is sufficient to find a training algorithm which succeeds in adequate time for this single architecture and training set.

On the contrary, we may be searching for an optimum architecture. Therefore we train several architectures with a varying number of layers and neurons, and afterwards use the architecture with minimum generalization error. For this purpose we need a training algorithm which works well for architectures with a different number of hidden layers and units, but the training set, and in particular the number of input neurons, is fixed.

Furthermore, the training algorithms which are commonly used for the empirical risk minimization are uniform. This means that the standard training algorithms do not take into account the special structure, the number of neurons and hidden layers of the architectures. Therefore it would be nice if the training complexity scales well with respect to all parameters , including the size of the architecture, the number of input neurons, and the size of the training set. If this general training problem, where several parameters are allowed to vary, turns out to be NP hard this gives us a theoretical motivation to keep certain parameters as small as possible when designing the training task. An increase in these parameters would lead to an excessive increase in the training time for the standard uniform learning algorithms.

The situations may be even more complicated than described above. One reasonable task is to search for a network with small empirical error in an entire family of architectures. For example, we may search for a network with small empirical error which may have an arbitrary architecture, where only the maximum number of neurons is limited in order to limit the structural risk. This question occurs if the training algorithms are allowed to modify the architecture and to insert or delete appropriate units, which is valid, for example, in cascade correlation, pruning methods, or an application of genetic algorithms to neural networks [20, 33, 40, 77, 92, 123]. Here one extreme situation is to allow the use of an arbitrary number of neurons. Since any pattern set can be implemented with a standard architecture and a number of neurons, which equals the size of the training set plus a constant number, the training problem becomes trivial. Of course, this result is useless for concrete training tasks because we cannot expect adequate generalization from such a network. On the contrary, it turns out in this chapter that the training problem is NP hard for tasks where the number of neurons is fixed and not correlated to the number of patterns. Of course, it is of special interest to see what happens between these two extreme positions. Concerning this question, one practically relevant case is to bound the number of neurons using the results from the PAC setting which guarantee adequate generalization.

In the following we mainly deal with a somewhat different problem, which is a decision problem correlated to the learning task in one of the above settings. Instead of finding an optimum architecture and adequate weights we only ask whether a network of the specified architecture and concrete weights exist such that the empirical error is small for this network. The difference is that for our task an algorithm answering only 'yes' or 'no' is sufficient, whereas a training task even asks for the concrete architecture and weights if the answer is 'yes'. Since this decision task is a simpler problem than the training task all results concerning the NP completeness transfer directly to the practically relevant training problem. The results we obtain which state that the decision problem can be solved polynomially in certain situations lead to a solution for the corresponding training problem, too, because the proof methods are constructive.

To summarize the above argumentation we ask one single question which we deal with in this chapter:

1. When is the following problem solvable in polynomial time: Decide whether a pattern set can be loaded correctly by a neural architecture with a specified architecture.

This question turns out to be NP complete if a fixed multilayer feed-forward architecture with perceptron activation and at least two nodes in the first hidden layer is considered where the input dimension is allowed to vary. The problem is NP complete even if the input dimension is fixed, but the number of neurons may vary in at least two hidden layers. Furthermore, it is NP complete if the number of neurons in the hidden layer is allowed to vary depending on the number of input patterns. But for any fixed folding architecture with perceptron activation the training problem is solvable in polynomial time without any restriction on the maximum input height.

In the third section we deal with the sigmoidal activation function instead of the perceptron function and prove that the classical loading problem to decide whether a pattern set can be classified correctly with a sigmoidal three node architecture is NP hard if two additional, but realistic restrictions are fulfilled by the architecture.

First the in principle loading problem will be defined.

5.1 The Loading Problem

The *loading problem* is the following task: Let $\mathcal{F} = \{F_l \mid l \in \mathbb{N}\}$ be a family of function classes which are represented in an adequate way. Given as an input a class $F_l \in \mathcal{F}$ and a finite set of points $P = \{(x_i, y_i) \mid i = 1, \ldots, m\}$, where F_l and P may have to fulfill some additional properties, decide whether a function $f \in F_l$ exists such that $f(x_i) = y_i$ for all i.

Since we are interested in the complexity of this problem, it is necessary to specify for any concrete loading problem the way in which F_l and P are to be represented. In tasks dealing with neural network learning, F_l often forms the set of functions that can be implemented by a neural architecture of a specified form. In this case we can specify F_l by the network graph. P is the training set for the architecture. It can be chosen in the following settings dealing with NP complexity as a set of vectors with elements in $\mathbb{Q}$, which are represented by some numerator and denominator without common factors in a decimal notation. Another possibility could be the assumption that the patterns consist of integers. This does not in fact affect the complexity of the problem if we deal with the perceptron activation function. Here any finite pattern set with rational input coefficients can be substituted by a pattern set with coefficients from $\mathbb{Z}$. This set requires the same space for a representation and can be loaded if and only if the original set can be loaded. It is obtained by just multiplying the coefficients with the smallest common denominator.

Now the simplest feed-forward neural network consists of only one computation neuron. One loading problem in this context is given if F_l equals the class of functions computed by an architecture with one computation unit, unspecified weights and biases, and input dimension l. If the architecture is used for classification tasks and is equipped with the perceptron activation function, then this loading problem, where P is taken as a pattern set in $\mathbb{Q}^n \times \{0, 1\}$, is solvable in polynomial time if the patterns are encoded in a decimal representation. It can be solved through linear programming [63, 65].

We may consider larger feed-forward networks. Assume these networks all have the same fixed architecture with a fixed input dimension and any computation unit has at most one successor unit. Then it is shown in [83] that the loading problem for such a function class is solvable in polynomial time, provided that the activation functions are piecewise polynomial function.

The situation changes if architectural parameters are allowed to vary. Judd considers the problem where F_l is given by the class of network functions implemented by a special feed-forward architecture with perceptron activation [62]. Here the number of computation and output neurons and the number of connections is allowed to vary if l varies. The NP completeness of this loading problem is proven by means of architectures which correspond to the SAT problem. The construction is valid even if the input dimension in any F_l is restricted to two, each neuron has at most three predecessors, and the network depth is limited by two. As a result, the construction of uniform learning algorithms which minimize the empirical error of arbitrary feed-forward architectures turns out to be a difficult task. However, the connection structure in the architectures that are considered in [62] is of a very special form and not likely to appear in practical applications. Furthermore, the output dimension is allowed to vary, whereas in practical applications it is usually restricted to only a few components.

A more realistic situation is addressed in [19]: It is shown that the loading problem is NP complete if F_l is the class of functions which are computed by a feed-forward perceptron network with l inputs, two hidden nodes, and only one output. The variation of the input dimension l seems appropriate to model realistic problems. A large number of features is available, for example, in image processing tasks [60, 88]. The features encode relevant information of the image but lead to an increase in the computational costs of the training, apart from the decrease in the generalization ability. The result of Blum and

Rivest is generalized to architectures with a fixed number $k \geq 2$ of hidden neurons and an output unit which computes an AND in [6]. Lin and Vitter show a similar result as Blum and Rivest for a cascade architecture with one hidden node [78]. In [108] the question is examined whether a restriction of the training set, such that the single patterns have limited overlap, and a restriction of the weights make the problems easier.

However, all these results deal with very restricted architectures and cannot be transformed to realistic architectures with more than one hidden layer directly. One reason for the difficulty of the loading problem can be given by the fact that the approximation capability of the function class is too weak. The use of a more powerful class possibly makes the loading problem easier. This problem is discussed in [109].

Another drawback of these results is that they deal with the perceptron activation function. In contrast, the most common activation function in concrete learning tasks is a sigmoidal function. Therefore much effort has been made to obtain results for a sigmoidal, or at least continuous activation, instead of the perceptron function. In [26] Dasgupta et.al. succeed in generalizing the NP result of Blum and Rivest to a three node architecture, where the hidden nodes have a semilinear activation. One price they pay is that the input biases are canceled.

Of course, we are particularly interested in results concerning the loading problem for sigmoidal architectures. Considering a fixed sigmoidal architecture it is expected that the loading problem is at least decidable [86]. The statement relies on the assumption that the so called Schanuel conjecture in number theory holds. But the complexity of training fixed networks is not exactly known in the sigmoidal case. One problem in this context is that polynomial bounds for the weights of a particular network which maps a finite data set correctly are not known.

When considering architectures with varying architectural parameters the sigmoidal function is addressed in [134]. Here the NP hardness of loading a three node network with varying input dimension, sigmoidal hidden nodes, and a hard limiter as output is proved provided that one additional so-called output separation condition is fulfilled. This condition is automatically fulfilled for any network with output bias 0. Unfortunately, canceling the output bias prohibits a transformation of the result from the standard sigmoidal function to the function tanh with the usual classification reference 0. Hoeffgen proves the NP completeness of the loading problem if $\mathcal{F}$ is an architecture with varying input dimension, two sigmoidal hidden nodes, and a linear output, which means that the network is used for interpolation instead of classification purposes. One crude restriction in this approach is that the weights have to be taken from $\{-1, 1\}$ [55].

Interesting results on sigmoidal networks are obtained if the question of simple classification is tightened to the task of an approximate interpolation or a minimization of the quadratic empirical error of the network. In this context, Jones proves that it is NP hard to train a sigmoidal three node network with varying input dimension and linear output with bias 0 such that the quadratic error on a training set is smaller than a given constant [61]. This argumentation even holds for more general activation functions in the hidden layer. This result is generalized to a sigmoidal architecture with $k \geq 2$ hidden nodes in [135].

Finally, modifying the problem of classification to the more difficult task of a minimization of the empirical error even adds some interesting aspects to the case we have started with, the simple perceptron unit. It is proved in [55] that it is not possible to find near optimum weights for a single perceptron in a real learning task in reasonable time unless RP = NP.

All results we have cited so far are stated for simple feed-forward networks, but obviously transfer directly to NP hardness results for recurrent and folding networks. Here training does not become easier since training feed-forward networks is a special case of training folding architectures. Therefore this subproblem is to be solved if we deal with the training of general folding networks, too. But when training a fixed folding architecture an additional parameter that may be allowed to vary in a function class $\mathcal{F}$ occurs: the maximum input height of a training set. In [27] the loading problem for fixed recurrent networks with linear activation where the input length is allowed to vary is considered and proved to be solvable in polynomial time.

The argument that the loading problem is easier than a training task, and therefore NP hardness results of the first problem transfer directly to the latter case, motivated us at the beginning of this chapter to mainly deal with the loading problem instead of considering practical learning tasks. Actually, the connection between training and loading can be made more precise as, for example, in [26, 56, 99]. First, the possibility of training efficiently in a learning task or to learn in polynomial time is defined formally. One possibility is the following modification of a definition in [99].

Definition 5.1.1. *Assume $\mathcal{F} = \bigcup_{n=1}^{\infty} F_l$ is a function class, where the input set of the functions in F_l is X_l and the output is contained in $Y = [0,1]$. $\mathcal{F}$ is polynomially learnable if an algorithm*

$$h : \bigcup_{l=1}^{\infty} \left(\{l\} \times \bigcup_{m=1}^{\infty} (X_l \times Y)^m \right) \to \mathcal{F}$$

exists with the following properties: $h(l, \_)$ or h^l for short maps to F_l. h runs in time polynomial in l and the representation of the inputs $(x_i, y_i)_i$. For all ϵ, $\delta > 0$ and $l \geq 1$ a number $m_0 = m_0(\epsilon, \delta, l)$ exists which is polynomial in l, $1/\epsilon$, and $1/\delta$ such that for all $m \geq m_0$ and for all probability distributions P on X_l and functions $f : X_l \to Y$ the inequality

$$P^m(\mathbf{x} \mid d_P(f, h_m^l(f, \mathbf{x})) - \min_{g \in F_l} d_P(f, g) > \epsilon) \leq \delta$$

is valid.

Actually, this notation only adds to the term of distribution-independent model-free PAC learnability the possibility of stratifying the function class and allowing a polynomial increase in the number of examples in accordance to this stratification – a definition which is useful if recursive architectures are considered as we have already seen, but which may be applied to uniform algorithms to train neural architectures with a varying number of inputs or hidden units as well. Furthermore, the limitation of the information theoretical complexity via the sample size m is accompanied by the requirement of polynomial running time of the algorithm h, as in the original setting of Valiant [129].

Note that sometimes a learning algorithm is defined such that it uses the parameters ϵ and δ, too [129]. This enables the algorithm to produce only a solution which is ϵ_1-consistent with some ϵ_1 sufficient to produce the actual accuracy, for example. In this case the algorithm is required to be also polynomial in $1/\epsilon$ and $1/\delta$. Even with this more general definition the next Theorem holds. The loading problem we deal with in the remainder of this chapter is not affected.

In analogy to the argumentation in [99] the following result can be shown:

Theorem 5.1.1. *Assume $\mathcal{F} = \{F_l \mid l \in \mathbb{N}\}$ is a family of function classes, where F_l maps X_l to $Y \subset \mathbb{R}$ with $|Y| < \infty$. Assume the representation of F_l takes at least space l. Assume the loading problem for $\mathcal{F}$ and arbitrary patterns P is NP hard. Then $\mathcal{F}$ is not polynomially learnable unless $RP = NP$.*

Proof. Assume $\mathcal{F}$ is polynomially learnable. Then we can define an RP algorithm for the loading problem as follows: For an instance (F_l, P) of the loading problem define U as the uniform distribution on the examples in P and $\epsilon = \epsilon_0/(1 + |P|)$, $\delta = 0.5$, where ϵ_0 is the minimum distance of two points in Y. The algorithm computes $m = m(\epsilon, \delta, l)$, chooses m examples according to U, and starts the polynomial learning algorithm h on this sample. The entire procedure runs in time polynomial in the representation length of F_l and P, and outputs with probability 0.5 a function f such that the values of f are ϵ-close to the values of any function in F_l approximating the pattern set P best. Because of the choice of ϵ the values of f are identical to the values of a best approximator of P. Hence we have either found with probability 0.5 a function in F_l which exactly coincides with the pattern set P if it exists and can answer 'yes', or we have unfortunately missed this function with probability 0.5, or it does not exist. In both cases we answer 'no', which is wrong with probability 0.5. This leads to an RP algorithm for the loading problem which contradicts the inequality RP$\neq$NP. $\qquad\square$

From this argumentation it follows that the class of three node perceptron networks stratified according to their input dimension, for example, is not polynomially learnable due to complexity problems. In contrast, the results which deal with the complexity of minimizing the quadratic empirical error limit the capability of special learning algorithms which try to minimize this error directly.

Since many common training algorithms for neural networks are based on some kind of gradient descent method, it is appropriate to consider the complexity of training with this special method in more detail. The examination of the drawbacks that occur when feed-forward networks are trained with gradient descent has led to remarkable improvements of the original algorithm [103, 141]. In particular, training networks with a large number of hidden layers is very difficult for simple back-propagation due to an exponential decrease in the error signals that are propagated through the layers. Several heuristics prevent the decrease in the signals and speed up the learning considerably.

Unfortunately, the same drawback can be found in recurrent and folding networks since they can be seen as feed-forward networks with a large number of hidden layers and shared weights. Here the same improved methods from the feed-forward case cannot be applied directly due to the weight sharing. Several other approaches have been proposed to overcome this difficulty, like learning based on an EM approach [14], some kind of very tricky normalization and recognition of the error signals in LSTM [54], or substituting an additional penalty term for the weight sharing property [67]. The detailed dynamics of gradient descent are studied, for example, in [16, 71]. Several approaches even characterize situations in which gradient descent behaves well in contrast to the general situation or try to find regions where the network behaves nicely, is stable, for example, and where training may be easier [17, 35, 79].

5.2 The Perceptron Case

In this section we consider only networks with the perceptron activation function. However, the architectures themselves we deal with, for example, in NP results, are realistic since they contain several hidden layers and an arbitrary number of hidden units.

5.2.1 Polynomial Situations

First of all it is possible to train any fixed feed-forward perceptron architecture in polynomial time. One training algorithm can be obtained directly by a recursive application of an algorithm which is proposed by Meggido for k-polyhedral separating of points [89], and which is already applied in [26] to a perceptron architecture with one hidden layer. The following is a precise formulation of an analogous algorithm for arbitrary feed-forward perceptron architectures.

Theorem 5.2.1. *For any fixed feed-forward architecture with perceptron activation function there exists a polynomial algorithm which chooses appropriate weights for a pattern set if it can be loaded and outputs 'no' otherwise.*

Proof. Enumerate the N neurons of the architecture in such a way that all predecessors of neuron j are contained in $\{1, \ldots, j-1\}$. Assume p points are to be classified by such an architecture. Each neuron in the network defines a hyperplane separating the input patterns of this neuron that occur during the computation of the p different outputs. We can assume that no input pattern of this neuron lies directly on the separating hyperplane. Therefore the behavior of the neuron is described by p inequalities of the form $\mathbf{w}^t \cdot \text{input}_\mathbf{p} + \theta \geq 1$ or $\mathbf{w}^t \cdot \text{input}_\mathbf{p} + \theta \leq -1$, where $\text{input}_\mathbf{p}$ is the input of the neuron when computing the output value of the entire network of point $\mathbf{p}$, $\mathbf{w}$ are the weights of this neuron, and θ is the bias. But as in [89] (Proposition 9/10), an equivalent hyperplane can be uniquely determined by the choice of at most $d+1$ of the above inequalities ($d =$ input dimension of the neuron) for which an exact equality holds instead of a $\leq$ or $\geq$. For each choice of possible equalities it is sufficient to solve a system of linear equations to find concrete weights for the neuron.

This leads to a recursive computation of the weights as follows: Assume $1, \ldots, n$ are the input neurons. Set $o_j^i = p_j^i$ for $j = 1, \ldots, n, i = 1, \ldots, p$, where $\mathbf{p}^i$ is the ith input pattern. Compute $\text{rec}(n+1, \{o_j^i | i = 1, \ldots, p, j = 1, \ldots, n\})$, where rec is the procedure:

```
rec(k, [o_j^i | i = 1, ..., p, j = 1, ..., k - 1]):
{
        Assume k_1, ..., k_l are the predecessors of neuron k.
        For all choices of ≤ l + 1 points of {(o_{k_1}^i, ..., o_{k_l}^i) | i = 1, ..., p}
        and a separation into positive and negative points:
            The choice defines ≤ l + 1 equalities w^t · o^i + θ = ±1
            for the parameters w and θ of neuron k.
            Compute the corresponding weight vector.
            Compute [o_k^i | i = 1, ..., p] as the image of (o_{k_1}^i, ..., o_{k_l}^i).
            If k = N and (o_{N-m}^i, ..., o_N^i) = y^i for all output neurons
            N - m, ..., N and all patterns y^i:
                Output the actual weights.
            If k ≠ N:
                Compute rec(k + 1, [o_j^i | i = 1, ..., p, j = 1, ..., k]).
}
```

This procedure outputs for all possible hyperplane settings one weight vector. The running time is limited by the product of the possible choices of at most $l+1$ of the p input points for any neuron and the separation of these points into two sets. It is limited by the term

$$\prod_{i=n+1}^{N} \binom{p}{l_i + 1} \cdot 3^{l_i+1} \leq (3p)^{(l_{n+1}+\ldots+l_N+N-n)},$$

where l_i is the number of predecessors of neuron i. $\qquad\square$

Unfortunately, this bound is exponential in the number of neurons and the number of predecessors of the single neurons. Therefore it does not lead to good bounds if architectural parameters are allowed to vary. Furthermore, it cannot be applied to recurrent architectures because here, the number N would increase with increasing input length. Fortunately, another argumentation shows that recurrence does not add too much complexity to the learning task:

Theorem 5.2.2. *For any fixed folding architecture with perceptron activation it can be decided in polynomial time whether a pattern set can be loaded correctly even if the input height is not restricted.*

Proof. Assume a folding architecture with perceptron activation function and an input set P are given where the number of different labels which occur in any tree in P adds up to L. Note that at a computation of the activation of one neuron which has i input neurons among its l predecessors the output value is uniquely determined by the value of the inputs – one of L^i possible input vectors – and the value of the other predecessors – one of 2^{l-i} possible vectors. The activation of any neuron with l predecessors without input neurons is determined by the value of the l predecessors – one of 2^l possible vectors.

If we write the different linear terms that may occur at the computation of the values on P to one vector then we obtain a vector of length $\sum_{i=1}^{N} 2^{l_i} L^{i_i}$, where N is the number of neurons, l_i is the number of predecessors of neuron i, which are computation or context neurons, and i_i is the number of predecessors which are input units. This is a vector of polynomial length containing linear polynomials in the weights if the architecture is fixed. We can determine the output of the network if the signs of each polynomial in this vector are known. We can check in polynomial time whether one sign vector corresponds to a computation in the folding architecture, which leads to a correct classification. (See Fig. 5.1 for an example.)

Therefore the following procedure can solve the loading problem: Find all possible sign vectors which occur if the polynomials are computed for some concrete weights. Afterwards, test whether at least one of these sign vectors describes a correct classification. One algorithm which finds all possible sign vectors of a vector of polynomials and which is polynomial in the length of the vector and the coefficients, but exponential in the number of parameters, is described in [11, 27]. Since the number of parameters is fixed in a fixed architecture the algorithm runs in polynomial time. $\qquad\square$

As a consequence, we can efficiently learn a perceptron architecture from a computational point of view if we fix the structure. This fact implies poly-

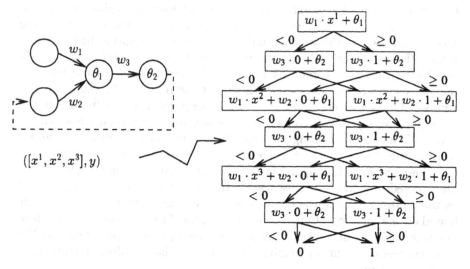

Fig. 5.1. Linear terms occurring at a computation of a concrete input for a folding architecture.

nomial learnability for feed-forward perceptron architectures with a fixed structure because the number of examples necessary for valid generalization increases only polynomially with $1/\epsilon$ and $1/\delta$. On the contrary, folding architectures with perceptron activation function do not lead to polynomial learnability because of information theoretical problems, as we have seen in the last chapter. But the complexity of learning does not increase in a more than polynomial way.

5.2.2 NP-Results

However, the question arises in the feed-forward as well as in the recurrent scenario as to how the complexity scales with respect to architectural parameters. The argumentation in Theorems 5.2.1 and 5.2.2 leads to bounds which are exponential in the number of neurons. Such a scaling prohibits polynomial learning in the sense of Definition 5.1.1, where the stratification is given by the number of network parameters: An algorithm as in Theorems 5.2.1 and 5.2.2 leads to an enormous increase in the computation time if the input class F_l becomes more complex.

But note that the loading problem is contained in NP for feed-forward as well as for recurrent architectures with the perceptron activation function since we can guess the weights – they are appropriately limited because they can be obtained as a solution of a system of linear equations as follows from Theorem 5.2.1 – and test whether these weights are correct.

In the following we derive several NP completeness results for the feed-forward case. Of course, the recurrent case is at least as complex as the feed-forward scenario, therefore the training of folding architectures is NP hard in comparable situations, too. Since we consider multilayer feed-forward architectures where the weights and biases are mostly unspecified, a somewhat different representation than the network graph is useful: We specify the function class F, computed by a certain feed-forward architecture, by the parameters $(n, n_1, \ldots, n_h, 1)$, where n is the number of inputs, h is the number of hidden layers, and $n_1, \ldots, n_h$ is the number of hidden units in the corresponding layer. We assume that the architecture has only one output neuron. In a concrete loading problem, $\mathcal{F}$ is defined by a family of such architectures where some of the parameters, for example, the input dimension n, are allowed to vary. It is appropriate to assume that the numbers $(n, n_1, \ldots)$ are denoted in a unary representation because the size n is a lower bound on the representation of the training set and the numbers $n_1, \ldots$ are a lower bound on the representation of concrete weights of such a network. Therefore any training algorithm takes at least $n_1 + \ldots + n_h$ time steps. Of course, NP hardness results for unary numbers are stronger than hardness results for a decimal notation.

Consequently, instances of any loading problem in this section have the form $(n, n_1, \ldots, n_h, 1)$, P, where the architecture is contained in a specified class $\mathcal{F}$ as mentioned above, and P is a finite training set with patterns in $\mathbb{Q}^n \times \{0, 1\}$ which is denoted in a decimal notation.

The result of Blum and Rivest [19] shows that an increase of the input dimension is critical and leads to NP completeness results. But the architecture considered in [6, 19, 108] is restricted to a single hidden layer with a restricted output unit. Here we generalize the argumentation to general multilayer feed-forward architectures.

Theorem 5.2.3. *Assume $\mathcal{F} = \{F_n \mid n \in \mathbb{N}\}$, where F_n is given by a feed-forward perceptron architecture $(n, n_1 \geq 2, n_2, \ldots, n_h, 1)$ with $h \geq 1$. In particular, only n is allowed to vary. Then the loading problem is NP complete.*

Proof. In [6] it is shown that for fixed $h = 1$ and $n_1 \geq 2$, but varying input dimension n and binary patterns, the loading problem is NP-complete if the output computes $o(x_1, \ldots, x_{n_1}) = x_1 \wedge \ldots \wedge x_{n_1}$ for $x_i \in \{0, 1\}$. We reduce this loading problem to the loading problem as stated in the theorem. One instance given by the architecture $(n, n_1, 1)$, n_1 fixed and $o = $ AND is mapped to an instance given by the architecture $(\tilde{n}, n_1, \ldots, n_h, 1)$ with fixed $h, n_2, \ldots, n_h$ as above, and $\tilde{n} = n + n_1 + 1$. Assume a pattern set $P = \{(\mathbf{x}_i, y_i) \in \mathbb{Q}^n \times \{0, 1\} \mid i = 1, \ldots, m\}$ and an architecture of the first type are given. We enlarge P to guarantee that a bigger architecture necessarily computes an AND in the layers following the first hidden layer. Define

$$
\begin{aligned}
\tilde{P} = \ & \{((\mathbf{x}_i, 0, \ldots, 0), y_i) \in \mathbb{R}^{\tilde{n}} \times \{0, 1\} \mid i = 1, \ldots, m\} \\
& \cup \{((0, \ldots, 0, \tilde{\mathbf{z}}_i, 1), 0), ((0, \ldots, 0, \tilde{\mathbf{z}}_i, 1), 1) \mid i = 1, \ldots, n_1(n_1 + 1)\} \\
& \cup \{((0, \ldots, 0, \mathbf{p}_i, 1), q_i) \mid i = 1, \ldots, 2^{n_1}\},
\end{aligned}
$$

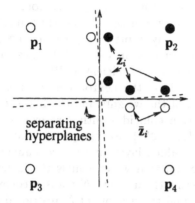

Fig. 5.2. Additional points in $\tilde{P}$ which determine the behavior of large parts of the neural architecture.

where $\tilde{z}_i, \bar{z}_i, p_i \in \mathbb{R}^{n_1}$, $q_i \in \{0, 1\}$ are constructed as follows:

Choose $n_1 + 1$ points with positive coefficients and zero ith coefficient on each hyperplane $H_i = \{x \in \mathbb{R}^{n_1} \mid$ the ith component of x is $0\}$, such that $n_1 + 1$ of these points lie on one hyperplane if and only if they lie on one H_i. Denote the points by $z_1, z_2, \ldots$ Define $\tilde{z}_i \in \mathbb{R}^{n_1}$ such that $\tilde{z}_i$ equals z_i, except for the ith component, which equals a small positive value ϵ. Define $\bar{z}_i$ in the same way, but with ith component $-\epsilon$. ϵ can be chosen such that if one hyperplane in $\mathbb{R}^{n_1}$ separates at least $n_1 + 1$ pairs $(\tilde{z}_i, \bar{z}_i)$, these pairs coincide with the $n_1 + 1$ pairs corresponding to the $n_1 + 1$ points on one hyperplane H_i, and the separating hyperplane nearly coincides with H_i (see Fig. 5.2). This is due to the fact that $n_1 + 1$ points z_i do not lie on one hyperplane if and only if $\det \begin{pmatrix} 1 & \cdots & 1 \\ z_1 & \cdots & z_{n_1+1} \end{pmatrix} \neq 0$. If ϵ is small enough the same inequality holds if z_i is substituted by any point on the line from $\tilde{z}_i$ to $\bar{z}_i$. p_i are all points in $\{-1, 1\}^{n_1}$, and $q_i = 1 \Leftrightarrow p_i = (1 \ldots 1)$. After decreasing the above ϵ if necessary we can assume that no p_i lies on any hyperplane which separates $n + 1$ pairs $\tilde{z}_i$ and $\bar{z}_i$.

Assume P can be loaded with a network N of the first architecture. We construct a solution of $\tilde{P}$ with a network $\tilde{N}$ of the second architecture: For each of the neurons in the first hidden layer of $\tilde{N}$ choose the first n weights and the bias as the weights or bias, respectively, of the corresponding neuron in N. The next n_1 weights are 0, except for the ith weight in the ith neuron of the first hidden layer of $\tilde{N}$ which is 1 and the $n + n_1 + 1$st weight which is $-\theta_i$ if θ_i is the bias of the ith neuron in the hidden layer of N. The neurons in the other layers of $\tilde{N}$ compute an AND. This network maps $\tilde{P}$ correctly.

Assume $\tilde{P}$ can be loaded with a network $\tilde{N}$ of the second architecture. The points $(\ldots, \tilde{z}_i, 1)$ and $(\ldots, \bar{z}_i, 1)$ are mapped differently, therefore each pair is separated by at least one of the hyperplanes defined by the neurons in the first hidden layer. Because $n_1(n_1 + 1)$ points are to be separated, the hyperplanes defined by these neurons nearly coincide in the dimensions $n + 1, \ldots, n + n_1$ with the hyperplanes H_i we have used in the construction of $\tilde{\mathbf{z}}_i$ and $\bar{\mathbf{z}}_i$, where the dimension $n + n_1 + 1$ serves as an additional bias for inputs with corresponding coefficient 1. We can assume that the point $(\ldots, \tilde{\mathbf{z}}_i, 1)$ is mapped to 1 by the neuron corresponding to H_i. Maybe we have to change the sign of the weights and the biases beforehand, assuming w.l.o.g. that no activation coincides with zero exactly. Then the values of $\{(0, \ldots, 0, \mathbf{p}_i, 1) \mid i = 1, \ldots, 2^{n_1}\}$ are mapped to the entire set $\{0, 1\}^{n_1}$ by the neurons of the first hidden layer, and the remaining part of the network necessarily computes a function which equals the logical function AND on these values. Consequently, the network N maps P correctly if the weights in the first hidden layer equal the weights of $\tilde{N}$ restricted to dimension n, and if the output unit computes an AND. $\qquad\square$

As a consequence of this result the complexity of training a realistic perceptron architecture increases rapidly if the number of input neurons increases. This fact gives us a theoretical motivation to design any concrete learning task in such a way that the input dimension becomes small. Apart from an improved generalization capability this leads to short training times.

However, this argumentation is not valid for every pattern set. In fact, we have used a special pattern set in the above proof where the patterns are highly correlated. The situation may be different if we restrict the patterns to binary or even more special pattern sets. For example, if the patterns are orthogonal the loading problem becomes trivial: Any orthogonal pattern set can be loaded. Therefore one often uses a unary instead of a binary encoding, although such an encoding increases the input dimension in practical applications [112]. This method may speed up the learning because the patterns become uncorrelated and need fewer hyperplanes, which means fewer computation units for a correct classification.

Another problem that occurs if we have fixed the input representation is choosing an appropriate architecture. One standard method is to try several different architectures and afterwards to choose the architecture with minimum generalization error. But this method needs an efficient training algorithm for network architectures with a varying number of hidden units in order to be efficient. Unfortunately, even here the loading problem is NP complete and may therefore lead to an increasing amount of time which is necessary to find an optimum architecture with the above method.

Theorem 5.2.4. *The loading problem for architectures from the set $\mathcal{F} = \{(n, n_1, \ldots, n_h, 1) \mid n_1, \ldots, n_h \in \mathbb{N}\}$ is NP-complete for any fixed $h \geq 2$ and $n \geq 2$.*

Proof. In [89] it is shown that the problem of deciding whether two sets of points R and Q in $\mathbb{Q}^2$ can be separated by k lines, *i.e.*, each pair $\mathbf{p} \in R$ and $\mathbf{q} \in Q$ lies on different sides of at least one line, is NP-complete. Consider the training set

$$P = \{(\mathbf{x}_i, y_i) \in \mathbb{Q}^2 \times \{0,1\} \mid (\mathbf{x}_i \in R \wedge y_i = 1) \vee (\mathbf{x}_i \in Q \wedge y_i = 0)\}$$

for two sets of points R and Q. P can be loaded by a network with structure $(2, k, |R|, 1)$ if and only if R and Q are separable:

Assume P can be loaded. The hidden nodes in the first hidden layer define k lines in $\mathbb{R}^2$. Each point $\mathbf{p} \in R$ is separated from each point $\mathbf{q} \in Q$ by at least one line because otherwise the corresponding patterns would be mapped to the same value.

Assume R and Q are separable by k lines. Define the weights of the neurons in the first hidden layer according to these lines. Let $R = \{\mathbf{p}_1, \ldots, \mathbf{p}_m\}$. Let the jth hidden unit in the second hidden layer compute $(x_1, \ldots, x_k) \mapsto (\neg)x_1 \wedge \ldots \wedge (\neg)x_k$, where the $\neg$ takes place at x_i if $\mathbf{p}_j$ lies on the negative side of the ith hyperplane. In particular, the unit maps $\mathbf{p}_j$ to 1 and $\mathbf{q}$ to 0 for all $\mathbf{q} \in Q$. Consequently, if the output unit computes an OR the pattern set P is mapped correctly.

This argumentation can be directly transferred to $h \geq 2$ and $n \geq 2$. $\square$

Obviously, the same problem becomes trivial if we restrict the inputs to binary patterns. One immediate consequence of the NP results is that we can find a finite pattern set for arbitrary large networks such that it cannot be implemented by the network. As a consequence, any kind of universal approximation argument for feed-forward networks necessarily has to take the number of patterns into account. The number of parameters necessary for an approximation depends in general on this parameter.

Here another interesting question arises: What is the complexity of training if we allow the number of parameters to vary, but only in dependence on the number of training points? As already mentioned, one extreme position is to allow a number of hidden units that equals the size of the training set plus a constant. Then the loading problem becomes trivial because any set can be implemented.

Other situations are considered in the following. One problem that plays a key role in the argumentation is the set splitting problem which is NP complete.

Definition 5.2.1. *The k-set splitting problem (k-SSP) is the following task: A set of points $C = \{c_1, \ldots, c_n\}$ and a set of subsets of these points $S = \{s_1, \ldots, s_m\}$ are given. Decide whether a decomposition of C into k subsets $S_1, \ldots, S_k$ exists such that every set s_i is split by this decomposition. In a formal notation: Does $S_1, \ldots, S_k \subset C$ exist with $S_i \cap S_j = \emptyset$ for all $i \neq j$ and $\bigcup_{i=1}^{k} S_i = C$ such that $s_i \not\subset S_j$ for all i and j?*

The k-set splitting problem is NP complete for any $k \geq 2$, and it remains NP complete for $k = 2$ if S is restricted such that any s_i contains exactly 3 elements [37].

The set splitting problem is reduced to a loading problem for a neural architecture in [19] showing the NP completeness of the latter task. We use it in the next theorem, too.

Theorem 5.2.5. *For $\mathcal{F} = \{(n, n_1, 1) \mid n, n_1 \in \mathbb{N}\}$ and instances $(n, n_1, 1)$, P with the restriction $|P| \leq 2 \cdot (n_1 + 1)^2$ the loading problem is NP complete.*

Proof. We reduce an instance $C = \{c_1, \ldots, c_n\}$, $S = \{s_1, \ldots, s_m\}$ with $|s_i| = 3$ for all i of the 2-SSP to an instance $(n+3, n+m, 1)$ of the loading problem, where the pattern set P consists of $2n^2 + m^2 + 3mn + n + 3m + 3$ points. In particular, $|P| \leq 2(n + m + 1)^2$.

The 2-SSP (C, S) has a solution if and only if the $k = n + m$-SSP

$$
\begin{aligned}
C' &= \{c_1, \ldots, c_n, c_{n+1}, \ldots, c_{n'=n+k-2}\}, \\
S' &= \{s_1, \ldots, s_m, s_{m+1}, \ldots, s_{m'=m+(n+k-3)(k-2)}\}
\end{aligned}
$$

is solvable, where $\{s_{m+i} \mid i = 1, \ldots, n' - n\} = \{\{c_j, c_k\} \mid j \in \{n+1, \ldots, n + k - 2\}, k \in \{1, \ldots, n + k - 2\} \setminus \{j\}\}$. Note that $|s_i| \leq 3$ for all i. The new points and sets in C' or S', respectively, ensure that $k - 2$ of the k-splitting sets have the form $\{c_i\}$, where $i > n$, and the remaining two sets correspond directly to a solution of the original 2-set splitting problem.

Such an instance of the k-set splitting problem is reduced to an instance of the $(n' + 3, k, 1)$ loading problem as follows: The training set P consists of two parts. One part corresponds to the SSP directly, the other part determines that the output unit computes an AND or a function which plays an equivalent role in a feed-forward perceptron network. The first part consists of the following points:

- The origin $(0, \ldots, 0)$ is mapped to 1,
- a vector $(0, \ldots, 0, 1, 0, \ldots, 0, 1, 0 \ldots, 0)$ for any set $s_i \in S'$ which equals 1 at the coefficient j if and only if $c_j \in s_i$ is mapped to 1,
- the unit vectors $(0, \ldots, 0, 1, 0 \ldots, 0)$ with an entry 1 at the place i for $i \in \{1, \ldots, n'\}$ are mapped to 0.

Consequently, we have constructed a training pattern corresponding to any point c_i and set s_j, respectively, in the first part of the pattern set. The patterns in the second part have an entry 0 at the first n' places, an entry 1 at the $n'+3$rd place in order to simulate an additional bias, and the remaining two components $n' + 1$, $n' + 2$ are constructed as follow:

Define the points $\mathbf{x}^{ij} = (4(i - 1) + j, j(i - 1) + 4((i - 2) + \ldots + 1))$ for $i \in \{1, \ldots, k\}$, $j \in \{1, 2, 3\}$. These $3k$ points have the property that if three of them lie on one line then we can find an i such that the three points coincide with $\mathbf{x}^{i1}$, $\mathbf{x}^{i2}$, and $\mathbf{x}^{i3}$. Now we divide each point into a pair $\mathbf{p}^{ij}$ and $\mathbf{n}^{ij}$ of points which are obtained by a slight shift of $\mathbf{x}^{ij}$ in a direction that is orthogonal to the line $[\mathbf{x}^{i1}, \mathbf{x}^{i3}]$ (see Fig. 5.3). Formally, $\mathbf{p}^{ij} = \mathbf{x}^{ij} + \epsilon \mathbf{n}_i$ and $\mathbf{n}^{ij} = \mathbf{x}^{ij} - \epsilon \mathbf{n}_i$, where $\mathbf{n}_i$ is a normal vector of the line $[\mathbf{x}^{i1}, \mathbf{x}^{i3}]$ with a positive second coefficient and ϵ is a small positive value. ϵ can be chosen such that the following holds:

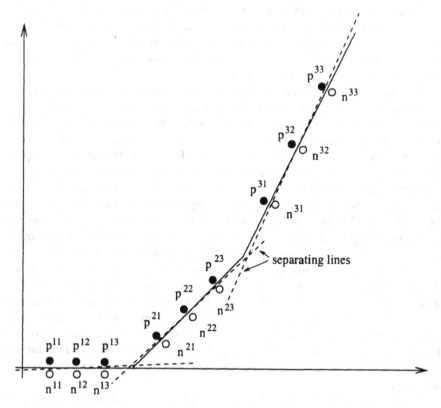

Fig. 5.3. Construction of the points $\mathbf{p}^{ij}$ and $\mathbf{n}^{ij}$: The points result by dividing each point x^{ij} on the lines into a pair.

Assume one line separates three pairs $(\mathbf{n}^{i_1j_1}, \mathbf{p}^{i_1j_1})$, $(\mathbf{n}^{i_2j_2}, \mathbf{p}^{i_2j_2})$, and $(\mathbf{n}^{i_3j_3}, \mathbf{p}^{i_3j_3})$, then the three pairs necessarily correspond to the three points on one line, which means $i_1 = i_2 = i_3$. This property is fulfilled if the determinant of all triples of points $(1, y_1^{ij}, y_2^{ij})$, where (y_1^{ij}, y_2^{ij}) is some point in $[\mathbf{x}^{ij} + \epsilon N, \mathbf{x}^{ij} - \epsilon N]$ and the indices i are different in such a triple, does not vanish. Using Proposition 6 of [89] it is sufficient for this purpose to choose $\epsilon \leq 1/(24 \cdot k(k-1) + 6)$ if N is a vector of length 1. Consequently, the representation of the points $\mathbf{n}^{ij}$ and $\mathbf{p}^{ij}$ is polynomial in n and m. The patterns $(0, \ldots, 0, p_1^{ij}, p_2^{ij}, 1)$ are mapped to 1, the patterns $(0, \ldots, 0, n_1^{ij}, n_2^{ij}, 1)$ are mapped to 0.

Now *assume that the SSP (S', C') is solvable* and let $S_1, \ldots, S_k$ be a solution. Then the corresponding loading problem can be solved with the following weights: The jth weight of neuron i in the hidden layer is chosen as $\begin{cases} -1 & \text{if } c_j \in S_i \\ 2 & \text{otherwise,} \end{cases}$ the bias is chosen as 0.5. The weights $(n'+1, n'+2, n'+3)$ of the ith neuron are chosen as $(-i+1, 1, -0.5+2 \cdot i(i-1))$ which corresponds to the line through the points $\mathbf{x}^{i1}$, $\mathbf{x}^{i2}$, and $\mathbf{x}^{i3}$. The output unit has the bias $-k+0.5$ and weights 1, $i.e.$, it computes an AND. With this choice of weights one can compute that all patterns are mapped correctly. Note that the point corresponding to $s_i \in C'$ is mapped correctly because s_i is not contained in any S_j and s_i contains at most 3 elements.

Assume, conversely, that a $solution$ of the $loading$ $problem$ is $given$. We can assume that the activations of the neurons do not exactly coincide with 0 when the outputs on P are computed. Consider the mapping which is defined by the network on the plane $\{(0,\ldots,0,x_{n+1},x_{n+2},1)\,|\,x_{n+1},x_{n+2} \in \mathbb{R}\}$. The points $\mathbf{p}^{ij}$ and $\mathbf{n}^{ij}$ are contained in this plane. Because of the different outputs each pair $(\mathbf{p}^{ij},\mathbf{n}^{ij})$ has to be separated by at least one line defined by the hidden neurons. A number $3k$ of such pairs exists. Therefore, each of the lines defined by the hidden neurons necessarily separates three pairs $(\mathbf{p}^{ij},\mathbf{n}^{ij})$ with $j \in \{1,2,3\}$ and nearly coincides with the line defined by $[\mathbf{x}^{i1},\mathbf{x}^{i3}]$. Denote the output weights of the network by $w_1,\ldots,w_k$ and the output bias by θ. We can assume that the ith neuron nearly coincides with the ith line and that the points $\mathbf{p}^{ij}$ are mapped by the neuron to the value 0. Otherwise we change all signs of the weights and the bias in neuron i, we change the sign of the weight w_i, and increase θ by w_i. But then the points $\mathbf{p}^{i2}$ are mapped to 0 by all hidden neurons, the points $\mathbf{n}^{i2}$ are mapped to 0 by all but one hidden neuron. This means that $\theta > 0$, $\theta + w_i < 0$ for all i and therefore $\theta + w_{i_1} + \ldots + w_{i_l} < 0$ for all $i_1,\ldots,i_l \in \{1,\ldots,k\}$ with $l \geq 1$. This means that the output unit computes the function $(x_1,\ldots,x_n) \mapsto \neg x_1 \wedge \ldots \wedge \neg x_n$ on binary values.

Define a solution of the SSP as $S_i = \{c_j \mid$ the jth unit vector is mapped to 1 by the ith hidden neuron$\}\backslash(S_1 \cup \ldots \cup S_{i-1})$. This definition leads to a decomposition of S into $S_1,\ldots, S_k$. Any set s_i is split because the situation in which all points corresponding to some c_j in s_i are mapped to 1 by one hidden neuron would lead to the consequence that the vector corresponding to s_i is mapped to 1 by the hidden neuron, too, because its bias is negative due to the classification of the zero vector. Then the pattern corresponding to s_i is mapped to 0 by the network.   □

If we are interested in an analogous result where the input dimension is fixed we can have a closer look at the NP completeness proof in [89] and obtain the following result:

Theorem 5.2.6. The $loading$ $problem$ for $architectures$ $from$ the set $\mathcal{F} = \{(2,n_1,n_2,1)\,|\,n_1,n_2 \in \mathbb{N}\}$ $with$ $instances$ $(2,n_1,n_2,1)$ and P $with$ $|P| \leq n_1^3$ as $input$ is NP $complete$.

Meggido reduces in [89] a 3-SAT problem with m clauses and n variables to a problem for k-polyhedral separability with $k = 8(u + 1) + nm$ lines and $m + nm^2 + 16(u+1)(m+nm^2)$ positive and $m + nm^2 + 32(u+1)(m+nm^2)$ negative points with coefficients that are polynomial in n and m and with a polynomial value u. This problem can be reduced as in the proof of Theorem

5.2.3 to a loading problem with the same number of points and $n_1 = k$, $n_2 = m + nm^2 + 16(u+1)(n + nm^2)$ hidden units. In particular, $(8(u+1) + nm)^3 \geq (m + nm^2)(2 + 16(u+1) + 32(u+1))$ for $n \geq 3$, $m \geq 3$.   □

One interesting question to ask is which of the above NP completeness results remain valid if we restrict the weights of a possible solution. This means an input for the loading problem does consist of the functions described by $(n, n_1, \ldots, n_h, 1)$, with the additional condition that in a network all weights and biases are absolutely bounded by some constant B. Here we assume that the weights are integers since for real or rational values, an arbitrary restriction would be possible by an appropriate scaling of the weights. This question is relevant since in practical applications a weight decay term is often minimized parallel to the empirical error minimization. It may be the case that training is NP complete, but only due to the fact that we sometimes have to find a solution with extreme weights which is hard to find. A weight restriction would lead to the answer 'no' in this case. Note that the formulations of the above theorems can be transferred directly to an analogous situation in which the weights are restricted. When considering the proofs, Theorem 5.2.3 holds with a weight restriction $B = k$. A slight modification of the proof allows the stronger bound $B = 2$: The SSP is NP complete even if the sets $s_i \in S$ contain at most 3 elements. Therefore, the biases of the first hidden layer are limited. Furthermore, we can substitute an AND by the function $x_1, \ldots, x_n \mapsto \neg x_1 \wedge \ldots \wedge \neg x_n$ in the network via a change of all weight signs in the first hidden layer. This function can be computed with weights and biases restricted by B.

Theorem 5.2.4 does not hold for restricted weights since a restriction of the weights would lead to a finite number of different separating hyperplanes in $\mathbb{R}^2$ which correspond to the neurons in the first hidden layer. Therefore a weight restriction reduces the loading problem to a finite problem which is trivial. The same is valid for Theorem 5.2.6.

In Theorem 5.2.5 the weights of a possible solution are unbounded in order to solve the more and more complicated situation which we have constructed in the last three input dimensions. But unfortunately, we are not aware of any argument which shows that the loading problem becomes easier for restricted weights in this situation.

We conclude this section with a remark on the complexity of algorithms which are allowed to change their architecture during the training process. If the changes are not limited then the training may produce a network with an arbitrary number of neurons. This does not generalize in the worst case but can approximate any pattern in an appropriate way: Here again, the loading problem becomes trivial.

In order to restrict the architecture in some way it is possible to formulate the learning task as the question as to whether an architecture with at most a limited number of neurons exists which loads the given data correctly. But the number of neurons does not necessarily coincide exactly with the maximum number for a concrete output. In a formal notation this corresponds to an input F_l of functions given by

$$\bigcup_{h^i \leq h, n_1^i \leq n_1, \ldots, n_{h^i}^i \leq n_{h^i}} (n, n_1^i, \ldots, n_{h^i}^i, 1)$$

instead of a simple vector $(n, n_1, \ldots, n_h, 1)$. We substitute in Theorems 5.2.3-5.2.6 the concrete input vector describing an architecture by the functions given by a network with at most these number of neurons. Then we obtain NP completeness results that cover algorithms which modify the architecture, too. The proofs of these theorems hold even for the slightly more general situation of an architecture in which the maximum number of neurons is restricted and the proofs need not be changed.

5.3 The Sigmoidal Case

Of course, the strongest drawback of the results in the previous section is that the activation function is a step function. In realistic applications one deals with a sigmoidal or at least continuous activation. In this section we derive one complexity result for the sigmoidal case which can be seen as a transfer of the complexity result for the 3-node perceptron architecture in [19] to the sigmoidal activation. In contrast to the work of [61, 135], we do not consider the task of minimizing the empirical error, but we use a sigmoidal network as a classifier. Since a classification is easier to obtain than an approximation, NP hardness results for a classification are more difficult to prove. In contrast to [134] we substitute the output separation condition by a condition which seems more natural and makes it possible to transfer the result to an arbitrary scaled or shifted version of the standard sigmoidal function, for example, the tanh. Furthermore, our result even holds for functions that are only similar to the sigmoidal function.

We deal with a feed-forward architecture of the form $(n, 2, 1)$ where the input dimension n is allowed to vary, the two hidden nodes have a sigmoidal activation σ instead of a perceptron activation, and the output function is the following modification of the perceptron activation

$$H_\epsilon(x) = \begin{cases} 0 & \text{if } x < -\epsilon, \\ \text{undefined} & \text{if } -\epsilon \leq x \leq \epsilon, \\ 1 & \text{otherwise}. \end{cases}$$

The purpose of this definition is to ensure that any classification is performed with a minimum accuracy. Output values that are too small are simply rejected. It is necessary to restrict the output weights, too, since otherwise any classification accuracy could be obtained by an appropriate scaling of the output values. Altogether this leads to a loading problem of the following form:

Definition 5.3.1. *The* loading problem *for a 3-node architecture with vary-ing input dimension, activation function* σ, *accuracy* $\epsilon > 0$, *and weight re-striction* $B > 0$ *is the loading problem given by* $\mathcal{F} = \{(n, 2, 1) \mid n \in \mathbb{N}\}$, *where the architecture* $(n, 2, 1)$ *has the activation function* σ *in the hidden nodes and activation function* H_ϵ *in the output node, and fulfills the additional con-dition that the output weights are absolutely bounded by* B. *The output bias may be arbitrary. The pattern set* P *can be any set in* $\mathbb{Q}^n \times \{0, 1\}$.

As already mentioned, the additional restrictions only ensure a certain ac-curacy of the classification. Now we can state the following theorem for a sigmoidal network.

Theorem 5.3.1. *The above loading problem given by* $\mathcal{F} = \{(n, 2, 1) \mid n \in \mathbb{N}\}$ *with activation function* $\sigma = \mathrm{sgd}$ *of the hidden nodes, output activation* H_ϵ, *and weight restriction* B *of the output weights is NP hard for any accuracy* $\epsilon \in\,]0, 0.5[$ *and weight restriction* $B \geq 2$.

Proof. First of all we take a closer look at the geometric situations which may occur in such a classification. A 3-node network computes the function

$$x \mapsto H_\epsilon(\alpha \,\mathrm{sgd}(\mathbf{a}^t\mathbf{x} + a_0) + \beta \,\mathrm{sgd}(\mathbf{b}^t\mathbf{x} + b_0) + \gamma)$$

for some weights α, β, γ, $\mathbf{a}$, $\mathbf{b}$, a_0, b_0, where $|\alpha| < B$ and $|\beta| < B$. As-sume a 3-node architecture classifies a pattern set correctly with accuracy ϵ and weight restriction B. The set of parameters such that the patterns are mapped correctly is an open set in $\mathbb{R}^{5+2n}$; therefore after a slight shift of the parameters, if necessary, we can assume that $\gamma \neq 0$, $\alpha + \gamma \neq 0$, $\beta + \gamma \neq 0$, $\alpha + \beta + \gamma \neq 0$, $\alpha \neq 0$, and $\beta \neq 0$. Furthermore, we can assume that $\mathbf{a}$ and $\mathbf{b}$ are linearly independent and the same is valid for all restrictions of $\mathbf{a}$ and $\mathbf{b}$ to at least two dimensions. We are interested in the boundary that is defined by

$$(*) \qquad \alpha \,\mathrm{sgd}(\mathbf{a}^t\mathbf{x} + a_0) + \beta \,\mathrm{sgd}(\mathbf{b}^t\mathbf{x} + b_0) + \gamma = 0\,.$$

This is empty or forms an $(n - 1)$-dimensional manifold M of the following form: If $\mathbf{x} \in M$, then $\mathbf{x} + \mathbf{v} \in M$ for any $\mathbf{v}$ orthogonal to $\mathbf{a}$ and $\mathbf{b}$. Conse-quently, M is constant in the directions orthogonal to $\mathbf{a}$ and $\mathbf{b}$; to describe M it is sufficient to describe the curve which is obtained if M is intersected with a plane containing $\mathbf{a}$ and $\mathbf{b}$ (see Fig. 5.4). Here we are only interested in the geometric form. Therefore, we can assume $\mathbf{a}^t\mathbf{x} + a_0 = x_1$, where x_1 is the first component of $\mathbf{x}$ after a rotation, translation, and scaling, if necessary, which does not affect the in principle form. Then the curve which describes M can be parameterized by x_1. A normal vector can be parameterized by $n(x_1) = \alpha \,\mathrm{sgd}'(x_1) \cdot \mathbf{a} + \beta \,\mathrm{sgd}'(\mathbf{b}^t\mathbf{x} + b_0) \cdot \mathbf{b}$, where the term $\mathrm{sgd}'(\mathbf{b}^t\mathbf{x} + b_0)$ can be substituted using $(*)$, which means

$$n(x_1) = \alpha \,\mathrm{sgd}'(x_1) \cdot \mathbf{a} + (-\gamma - \alpha \,\mathrm{sgd}(x_1)) \left(1 - \frac{-\gamma - \alpha \,\mathrm{sgd}(x_1)}{\beta}\right) \cdot \mathbf{b}\,.$$

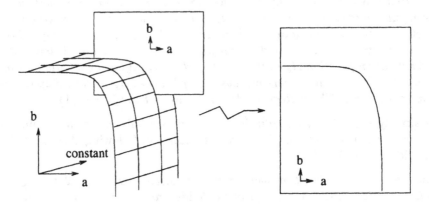

Fig. 5.4. Manifold M which forms the boundary of the classification: M can be entirely determined by the curve which is obtained if M is intersected with a plane containing the vectors **a** and **b**.

Define $\tilde{n}(x_1) = n(x_1)/|n(x_1)|$. Considering the four values γ, $\gamma + \alpha$, $\gamma + \beta$, and $\gamma + \alpha + \beta$ several cases result:

- *All values are positive or all values are negative:* M is empty.
- *One value is positive, the other three are negative:*
  Since $\text{sgd}(-x) = 1 - \text{sgd}(x)$ we can assume that $\gamma > 0$. Maybe we have to change the sign of the weights and the biases beforehand. If, for example, $\alpha + \gamma$ is positive, we substitute α by $-\alpha$, γ by $\alpha + \gamma$, and the weights of the first hidden neuron by their negative values without changing the entire mapping.
  Consequently, $\alpha < -\gamma$, and $\beta < -\gamma$. Dividing (∗) by γ we obtain $\gamma = 1$, $\alpha < -1$, and $\beta < -1$. The curve describing M looks like that depicted in Fig. 5.5, in particular, the region classified positively is convex, as can be seen as follows: For $\text{sgd}(x_1) \approx -1/\alpha$ the normal vector is $\tilde{n}(x_1) \approx -\mathbf{a}/|\mathbf{a}|$. For $\text{sgd}(x_1) \approx 0$ it is $\tilde{n}(x_1) \approx -\mathbf{b}/|\mathbf{b}|$. In general, $\tilde{n}(x_1) = \lambda_1(x_1)\,\mathbf{a} + \lambda_2(x_1)\,\mathbf{b}$ for appropriate functions λ_1 and λ_2. Assume that the curve is not convex. Then there would exist at least two points on the curve with identical $\tilde{n}$, identical λ_1/λ_2 and, consequently, at least one point x_1 with $(\lambda_1/\lambda_2)'(x_1) = 0$. But one can compute $(\lambda_1/\lambda_2)'(x_1) = C(x_1) \cdot (-\beta - 1 + \text{sgd}(x_1)(2\beta + 2) + \text{sgd}^2(x_1)(2\alpha + \alpha^2 + \alpha\beta))$ with some factor $C(x_1) = \alpha\beta\text{sgd}'(x_1)/((-1 - \alpha\text{sgd}(x_1))^2(1 + \beta + \alpha\text{sgd}(x_1))^2) \neq 0$. If $(\lambda_1/\lambda_2)'(x_1)$ was 0, $\alpha = \beta = -1$ or

$$(**) \qquad \text{sgd}(x_1) = \frac{-\beta - 1}{\alpha(\alpha + \beta + 2)} \pm \sqrt{\frac{(1 + \beta)((\alpha + 1)^2 + \beta(\alpha + 1))}{\alpha^2(\alpha + \beta + 2)^2}},$$

where the term the square root is taken from is negative except for $\alpha = -1$ or $\beta = -1$ because $(1 + \alpha)$, $(1 + \beta)$, and $(1 + \alpha + \beta)$ are negative.

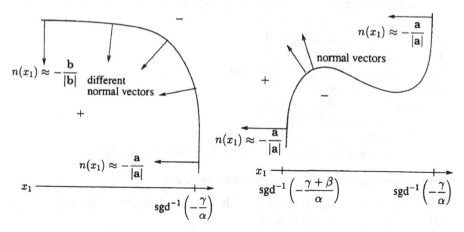

Fig. 5.5. Left: classification in case 2, precisely 1 value is positive; right: classification in case 3, precisely 2 values are positive.

– *Exactly two values are positive:*
We can assume that the positive values are γ and $\beta + \gamma$. Maybe we have to change the role of the two hidden nodes or the signs of the weights and the biases beforehand. Note that γ and $\alpha + \beta + \gamma$ cannot both be positive in this situation. If $\text{sgd}(x_1) \approx -\gamma/\alpha$ or $\text{sgd}(x_1) \approx (-\gamma - \beta)/\alpha$ then it holds that $\tilde{n}(x_1) \approx -a/|a|$. Arguing as before, we can assume $\gamma = 1$, $\alpha < -1$, $\beta > -1$, and $\alpha + \beta < -1$. The curve describing M has an S-shaped form (see Fig. 5.5) because there exists at most one point on the curve where $(\lambda_1/\lambda_2)'(x_1)$ vanishes: This point is $\text{sgd}(x_1) = 0.5$ if $\alpha + \beta + 2 = 0$, the point is the solution (∗∗) with the positive sign if $\alpha + \beta + 2 < 0$, and the solution (∗∗) with the negative sign if $\alpha + \beta + 2 > 0$.
– *Exactly 3 values are positive:* This case is dual to case 2.

To summarize, the classification can have one of the following forms:

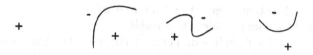

Now we reduce the 2-set splitting problem to a loading problem for this architecture.
Reduction: For an SSP (C, S) with $C = \{c_1, \ldots, c_n\}$ and $S = \{s_1, \ldots, s_m\}$, where we can assume $|s_i| = 3$ for all i [37], the following $m + n + 15$ patterns in $\mathbb{R}^{n+5}$ can be loaded if and only if the SSP is solvable:
Positive examples, *i.e.*, the output is to be 1, are

$(0, .., 1, .., 1, .., 1, .., 0, 0, 0, 0, 0, 0)$
 with an entry 1 at the place i, k, and l for any $s_j = \{c_i, c_k, c_l\}$ in S,

$(0, ..., 0, 0, 0, 0, 0, 0)$,
$(0, ..., 0, 1, 1, 0, 0, 0)$,
$(0, ..., 0, 0, 1, 1, 0, 0)$,
$(0, ..., 0, 0, 0, 0, -0.5, 0.5)$,
$(0, ..., 0, 0, 0, 0, 0.5, 0.5)$,
$(0, ..., 0, 0, 0, 0, c, c)$,
$(0, ..., 0, 0, 0, 0, -c, c)$,

 c being constant with
$$c > 1 + 4B/\epsilon \cdot \left(\mathrm{sgd}^{-1} \left(1 - \epsilon/(2B)\right) - \mathrm{sgd}^{-1} \left(\epsilon/(2B)\right)\right).$$

Negative examples, *i.e.*, the output is to be 0, are

$(0, ..., 1, ..., 0, 0, 0, 0, 0, 0)$ with an entry 1 at place i for $i \le n$,
$(0, ..., 0, 1, 0, 0, 0, 0)$,
$(0, ..., 0, 0, 1, 0, 0, 0)$,
$(0, ..., 0, 0, 0, 1, 0, 0)$,
$(0, ..., 0, 1, 1, 1, 0, 0)$,
$(0, ..., 0, 0, 0, 0, -1.5, 0.5)$,
$(0, ..., 0, 0, 0, 0, 1.5, 0.5)$,
$(0, ..., 0, 0, 0, 0, 1 + c, c)$,
$(0, ..., 0, 0, 0, 0, -1 - c, c)$ with c as above.

Assume that the SSP is solvable with a partition $C = S_1 \cup S_2$ where $S_1 \cap S_2 = \emptyset$. Consider the weights $\alpha = \beta = -1$, $\gamma = 0.5$,

$$\mathbf{a} = k \cdot (a_1, \ldots, a_n, 1, -1, 1, 1, -1)$$
$$\mathbf{b} = k \cdot (b_1, \ldots, b_n, -1, 1, -1, -1, -1),$$

$a_0 = -0.5 \cdot k$, $b_0 = -0.5 \cdot k$, where k is a positive constant and

$$a_i = \begin{cases} 1 & \text{if } c_i \in S_1 \\ -2 & \text{otherwise} \end{cases} \quad \text{and} \quad b_i = \begin{cases} 1 & \text{if } c_i \in S_2 \\ -2 & \text{otherwise} \end{cases}.$$

For appropriate k, this solves the loading problem with accuracy $\epsilon < 0.5$ because $\mathrm{sgd}(-x) \to 0$ and $\mathrm{sgd}(x) \to 1$ for $x \to \infty$.
Assume that the loading problem is solvable.
First, cases 3 and 4 are excluded. Then a solution of the SSP is constructed using the convexity of the positive region in case 2. Obviously, case 1 can be excluded directly.
Assume the classification is of case 3:
We only consider the last two dimensions, where the following problem is included: (We drop the first $n + 3$ coefficients which are 0.) The points $(-0.5, 0.5)$, $(0.5, 0.5)$, (c, c), $(-c, c)$ are mapped to 1 and the points $(-1.5, 0.5)$, $(1.5, 0.5)$, $(1 + c, c)$, $(-1 - c, c)$ are mapped to 0 (see Fig. 5.6a). Define $p_0 := \mathrm{sgd}^{-1}(\epsilon/(2B))$ and $p_1 := \mathrm{sgd}^{-1}(1 - \epsilon/(2B))$. $\{\mathbf{x} \mid p_0 \le \mathbf{a}^t\mathbf{x} + a_0 \le p_1\}$

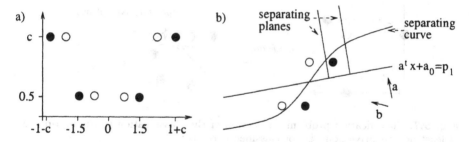

Fig. 5.6. a) Classification problem in the last two dimensions; b) outside the b-relevant region.

and $\{\mathbf{x} | p_0 \leq \mathbf{b}^t\mathbf{x} + b_0 \leq p_1\}$ are called the a- or b-relevant region, respectively. Outside, $\text{sgd}(\mathbf{a}^t\mathbf{x} + a_0)$ or $\text{sgd}(\mathbf{b}^t\mathbf{x} + b_0)$, respectively, can be substituted by a constant, the difference of the output is at most $\epsilon/2$.

Now the argumentation proceeds as follows: First it is shown that three points forming a triangle are contained in the a-relevant region. This leads to a bound for the absolute value of $\mathbf{a}$. Second, if three points forming a triangle are contained in the b-relevant region the same argumentation leads to a bound for the absolute value of $\mathbf{b}$. Using these bounds it can be seen that neighboring points cannot be classified differently. Third, if no such triangle is contained in the b-relevant region, the part $\mathbf{b}^t\mathbf{x} + b_0$ does not contribute to the classification of neighboring points outside the b-relevant region, the two points cannot be classified differently.

First step: Since the points with second component 0.5 cannot be separated by one hyperplane, one point $(x, 0.5)$ with $x \in [-1.5, 1.5]$ exists inside the a- and b-relevant region, respectively. If the points (c, c) and $(1 + c, c)$ were both outside the a-relevant region then they would be separated by any hyperplane with normal vector $\mathbf{b}$ which intersects the separating manifold outside the a-relevant region because the part given by $\mathbf{a}$ does not contribute to the classification (see Fig. 5.6b). The normal vector of the manifold is approximately $-\mathbf{a}/|\mathbf{a}|$ for large and small $\mathbf{b}^t\mathbf{x} + b_0$, respectively. Therefore we can find a hyperplane where both points are located on the same side. Contradiction. The same argumentation holds for $(-c, c)$ and $(-1 - c, c)$. Therefore the diameter of the a-relevant region restricted to the last two dimensions is at least $c - 1$. Consequently, $a \leq (p_1 - p_0)/(c - 1) < \epsilon/(4B)$, where $a = |(a_{n+4}, a_{n+5})|$.

Second step: If one of the points (c, c) and $(1 + c, c)$ and one of the points $(-c, c)$ and $(-1 - c, c)$ is contained in the b-relevant region, it follows that $b \leq \epsilon/(4B)$ for $b = |(b_{n+4}, b_{n+5})|$. This leads to a contradiction: For the points $\mathbf{x}_1 = (0, \ldots, 0, c, c)$ and $\mathbf{x}_2 = (0, \ldots, 0, 1 + c, c)$ we can compute

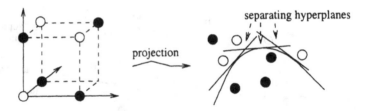

Fig. 5.7. Classification problem; projection of the classification to the a/b-plane, at least one negative point is not classified correctly

$$|\alpha \operatorname{sgd}(\mathbf{a}^t \mathbf{x}_1 + a_0) + \beta \operatorname{sgd}(\mathbf{b}^t \mathbf{x}_1 + b_0) + \gamma$$
$$-\alpha \operatorname{sgd}(\mathbf{a}^t \mathbf{x}_2 + a_0) - \beta \operatorname{sgd}(\mathbf{b}^t \mathbf{x}_1 + b_0) - \gamma|$$
$$\leq |\alpha| |\mathbf{a}^t \mathbf{x}_1 - \mathbf{a}^t \mathbf{x}_2| + |\beta| |\mathbf{b}^t \mathbf{x}_1 - \mathbf{b}^t \mathbf{x}_2| \leq \epsilon.$$

Because $|\alpha|, |\beta| \leq B$ and $|\operatorname{sgd}(x) - \operatorname{sgd}(x + \delta)| \leq \cdot \delta$ for $\delta > 0$.

Third step: If both points (c, c) and $(1 + c, c)$ or both points $(-c, c)$ and $(-1 - c, c)$ are outside the b-relevant region, the difference of the values $\operatorname{sgd}(\mathbf{b}^t \mathbf{x} - b_0)$ with corresponding $\mathbf{x}$ is at most $\epsilon/(2B)$. The same contradiction results.

Assume the classification is of case 4:
The classification includes in the dimensions $n + 1$ to $n + 3$ the problem depicted in Fig. 5.7. The negative points are contained in a convex region, each positive point is separated by at least one tangential hyperplane of the separating manifold M from all negative points. Consider the projection to a plane parallel to **a** and **b**. Following the convex curve which describes M the signs of the coefficients of a normal vector can change at most once. But a normal vector oriented in such a way that it points to the positive points which separates a positive point necessarily has the signs $(+, +, -)$ for $(1, 1, 0)$, $(-, +, +)$ for $(0, 1, 1)$, and $(-, -, -)$ for $(0, 0, 0)$ in the dimensions $n + 1$ to $n + 3$. Contradiction.

Solution of the SSP:
Consequently, the classification is of case 2. We can assume $\gamma = -1$, $\alpha > 1$, and $\beta > 1$. Define $S_1 = \{c_i \mid a_i \text{ is positive}\}$, $S_2 = C \backslash S_1$. Assume $s_l = \{c_i, c_j, c_k\}$ exists such that all three coefficients a_i, a_j, and a_k are positive. In the components i, j, k the classification $(1, 0, 0)$, $(0, 1, 0)$, $(0, 0, 1) \mapsto 0$ and $(0, 0, 0)$, $(1, 1, 1) \mapsto 1$ is contained. The positive points are contained in a convex region, each negative point is separated by at least one tangential hyperplane of the separating manifold M. We project to a plane parallel to **a** and **b**. Following the curve which describes M, the normal vector, oriented to the positive region, is $\approx -\mathbf{a}/|\mathbf{a}|$, then the sign of each component of the normal vector changes at most one time, finally it is $\approx -\mathbf{b}/|\mathbf{b}|$. But a vector where the three signs in dimension i, j, and k are equal cannot separate a negative point because $(0, 0, 0)$ and $(1, 1, 1)$ are mapped to 1. Furthermore,

the sign in dimension i has to be negative if c_i is separated, and the same is valid for j and k. Contradiction.

The same argumentation shows that not all three coefficients a_i, a_j, and a_k can be negative. □

Note that we have not used the special form of the sigmoidal function, but only some of its properties. Consequently, the same result with perhaps different accuracy ϵ and weight restriction B is valid for any activation function $\sigma : \mathbb{R} \to \mathbb{R}$ which is continuous, piecewise continuously differentiable, symmetric, squashing, and where the boundary limits a convex region in cases 2 and 4.

Furthermore, the request for a certain accuracy enables us to generalize the NP hardness result to functions which can be approximated by the sigmoidal function, but are not identical.

Corollary 5.3.1. *The loading problem for the 3-node architecture with accuracy $\epsilon \in]0, 1/3[$, weight restriction 2, and varying input dimension is NP-hard for any activation function σ which can be approximated by the standard sigmoidal activation in the following way: For all $x \in \mathbb{R}$, $|\sigma(x) - \mathrm{sgd}(x)| < \epsilon/8$ holds.*

Proof. Consider the reduction in the main proof with the function sgd, weight restriction 2, and accuracy $\epsilon/2$. If the SSP is solvable we can not only find a solution of the corresponding loading problem with accuracy $\epsilon/2$, but accuracy $3\epsilon/2 < 0.5$. This leads to a solution of the loading problem for the activation σ with weight restriction 2 and accuracy ϵ.

Conversely, any solution of the loading problem with activation σ, weight restriction 2, and accuracy ϵ can be transformed into a solution with activation sgd, weight restriction 2, and accuracy $\epsilon/2$. This leads to a solution of the SSP. □

Corollary 5.3.2. *The loading problem for the 3-node architecture with accuracy $\epsilon \in]0, 0.5[$, weight restriction $B > 0$, and varying input dimension is NP-hard for any activation function σ which can be written as $a\sigma(bx + c) + d = \mathrm{sgd}(x)$ for all $x \in \mathbb{R}$ and real numbers a, b, c, and d with $|a| \le B/2$.*

Proof. Consider a reduction of the SSP to a loading problem with accuracy ϵ and weight restriction $B/|a| \ge 2$ for the activation sgd. A solution for the activation function sgd with weight restriction $B/|a|$ directly corresponds to a solution for the activation function σ with weight restriction B. □

In particular, this result holds for any function which is a shifted or scaled version of the sigmoidal function like the hyperbolic tangent. Note that the concrete choice of ϵ and B in our main proof is not the only possibility. An increase of B leads to the possibility of increasing ϵ, too. In fact, any value $B \ge 2$ and $\epsilon \le B/4$ is a possible choice. This leads to corresponding values for ϵ and B in Corollary 5.3.1, too, but it does not affect the distance of σ and sgd which is at most $1/24$.

5.4 Discussion and Open Questions

In this chapter we have examined the complexity of training folding networks in a very rough setting. This means that we have first figured out several situations which seem of practical relevance: The training of a fixed architecture or the training of architectures where architectural parameters are allowed to vary in order to get universal training algorithms for these situations. Then we have focused on the question as to whether this task can be solved in polynomial time or is NP hard. We have substituted a correlated decision problem, the loading problem, for the learning problem. It has been pointed out that the NP hardness of this decision problem leads to a complexity theoretical barrier to learning such architectures polynomially unless $RP = NP$.

It has been shown that the training of a fixed folding architecture can be done in polynomial time if the activation function is the perceptron activation. For the standard sigmoidal function this is an unsolved problem.

When architectural parameters vary it has been shown in the perceptron case that an increasing number of input neurons may lead to an enormous increase in the training time for standard multilayer feed-forward perceptron networks. As a practical consequence any learning task should be designed in such a way that the number of inputs is small. Although this generalizes the result of [19] to realistic architectures, we have in contrast to [19] considered arbitrary patterns which are not necessarily binary patterns. The complexity of training may be much smaller if the patterns only have a binary form or – this seems even more promising – if the patterns have limited correlation. An attempt in this direction is made, for example, in [108]. Furthermore, our proof does not transfer to neural architectures with general connections. Since more general connection structures like additional direct links from the input units to the outputs, for example, are used in practical applications, results concerning a general structure would be interesting.

The consideration of arbitrary patterns instead of binary ones leads to the question of the complexity of training networks with varying architectures (compare Problem 12.13 in [132]). We have shown the NP completeness of this question, too. Unfortunately, we have used at least two hidden layers where the number of neurons varies. The problem which deals with the complexity of training a single hidden layer architecture where the number of neurons may vary remains unsolved.

We have considered situations which take into account training procedures that correlate the number of neurons and patterns or that may manipulate the architecture during the training via pruning or insertion of neurons, too. In this context we have proved several NP completeness results dealing with realistic architectures. Note that NP completeness results for architectures where the number of neurons and patterns are correlated are of special interest for concrete bounds on the number of neurons necessary for an interpolation of a finite set of points. The situations where NP completeness can be proven lead to lower bounds for the approximation scenario. Here it would be nice to modify the theorems in such a way that the number of neurons is allowed to increase linearly in the number of patterns.

When considering sigmoidal networks it does not seem likely that training becomes easier. But note that there is no theoretical motivation for this assumption and the complexity results do not transfer directly to the sigmoidal case. In fact, using the sigmoidal function instead of a step activation has enabled the training of networks via back-propagation [32, 136] in practice, whereas the lack of a reasonable training procedure for perceptron networks with more than one computation unit several years earlier had significantly reduced the interest in neural networks [90]. Therefore a theoretical investigation of the sigmoidal case is necessary, too.

Our NP results in the sigmoidal case are unfortunately restricted to the 3-node architecture and use arbitrary, not necessarily binary inputs. However, our modification of the classification task such that a minimum accuracy is guaranteed makes it possible to transfer the result to functions which nearly coincide with the standard sigmoidal function. Pathological examples of activation functions where a hidden oscillation of the function leads to a large capacity as in [119] are in some way excluded since nearly invisible oscillations can only have an effect if they are either reinforced with large weights or arbitrarily small outputs are allowed – both possibilities are excluded in our case.

Of course, it would be nice to obtain NP results for realistic sigmoidal architectures. Furthermore, a precise analytic characterization of the activation functions for which such an NP result holds would be interesting. Finally, results whether the training of sigmoidal architectures is contained in NP are missing because polynomial bounds on the weights are not known in contrast to the perceptron case.

Despite these NP results an investigation of the complexity of the concrete learning algorithms used for the training of recurrent and folding networks is necessary. Here it is adequate to consider the number of training steps that are used in the worst case or in general settings, and furthermore, the complexity of one single step has to be taken into account [98, 107, 139]. An analysis of numerical problems that may occur are interesting, in particular when dealing with recurrent or folding networks, and may give rise to the preference of some learning algorithms like LSTM [54] compared to others, *e.g.*, the simple gradient descent method.

Chapter 6

Conclusion

In this volume, folding networks have been examined which form a very promising approach concerning the integration of symbolic and subsymbolic learning methods. It has been proven that they are suitable as a learning method in principle. For this purpose, we have investigated the approximation capability, the learnability, and the complexity of learning for folding architectures. Apart from the argumentation specific for folding networks, we have obtained results that are interesting for conventional recurrent networks, standard feed-forward networks, or learning theory as well.

In the first part of the mathematical investigation, folding networks have been shown to be universal approximators if a measurable mapping is to be approximated in probability. This transfers to recurrent networks. In both cases, bounds on the number of neurons are given if a finite number of examples are to be interpolated. On the contrary, several restrictions exist if a mapping is to be approximated in the maximum norm with a recurrent or folding network. However, we have shown that as a computational model, recurrent networks with the standard sigmoidal activation function can compute any mapping on off-line inputs in exponential time.

Concerning learnability we have first contributed several results to the topic of distribution-dependent learnability: In [132] the term PUAC learnability is introduced as a uniform version of PAC learnability. We have shown that this is a different concept to PAC learnability, answering Problem 12.4 of [132]. In analogy to consistent PUAC learnability in [132] we have introduced the term of consistent PAC learnability and scale sensitive versions of these terms. We have established characterizations of these properties which do not refer to the notion of a learning algorithm, and are therefore of interest if the property is to be proved for a concrete function class. Additionally, we have examined the relations between these terms. We have discussed at which level learning via an encoding of an entire real valued function into a single value is no longer possible. Usually, learnability in any of the above formalisms is guaranteed by means of the finiteness of the capacity of a function class. Since it has turned out that the capacity of folding architectures with arbitrary inputs is in some sense unlimited, it has been necessary to take a closer look at the situation. For this purpose we have generalized two approaches which guarantee learnability even in the case of infinite capacity to function classes.

Since the capacity of a function class plays a key role in the generalization ability, we then estimated the capacity of folding architectures with restricted inputs. For recurrent networks, upper bounds already exist in the literature but the lower bounds in the literature deal with the pseudodimension, the finiteness of which is a sufficient but not necessary condition for learnability. We have substituted these bounds by bounds for the fat shattering dimension, a measure characterizing learnability under realistic conditions. In all cases the lower bounds depend on the maximum input height. Consequently, distribution-independent learnability cannot be guaranteed in general. But if the probability of high trees or the maximum height in a training set is limited, guarantees for correct generalization can be found. However, we have constructed a situation in which the number of examples necessary for valid generalization increases at least in a more than polynomial way with the required accuracy, which answers in particular Problem 12.6 of [132].

When dealing with the complexity of learning we have shown that efficient learning is possible for a fixed folding architecture with perceptron activation. All other results have shown the NP hardness of the loading problem – a problem which is even more simple than training – for a feed-forward architecture if several architectural parameters are allowed to vary. Since the training of a folding architecture is not easier, the results transfer directly to this case, too, but they are interesting concerning general neural network learning as well. For the perceptron activation function we have generalized the classical NP completeness result of Blum and Rivest [19] to realistic multilayer feed-forward architectures. We have considered the case where the number of hidden neurons is allowed to increase depending on the number of training examples. A result addressing Problem 12.13 in [132] has been added which is interesting if an optimal architecture is to be found for a concrete learning task. Furthermore, a generalization of the loading problem in [19] under two additional realistic conditions to the standard sigmoidal activation function has been presented, compare Problem 12.12 in [132]. All these results are of interest for neural network learning in principle, since they show which quantities should be kept small in the design of a learning problem, and which quantities do not slow down the training process too much.

Of course, several problems concerning the above topics remain open. We have already listed some of them in the conclusions of the respective chapters. Two problems seem of particular interest: Although we have obtained guarantees for the learning capability of folding architectures from an information theoretical point of view, the argumentation refers to the fact that inputs with an extensive recurrence become less probable. This is in some way contradictory to the original idea of recurrent or folding networks, since one advantage of these approaches is that they can deal with input data of *a priori* unlimited length or height, *i.e.*, with some kind of infiniteness. Therefore it would be interesting to see whether a guarantee for learnability can be found employing a specific property of the network instead, for example a stability criterion. Additionally, such a criterion may be useful for the efficiency of the learning process itself. A reduction of the search space for the weights to well behaved regions may prohibit an instability of the learning algorithm.

Another interesting question is the complexity of training sigmoidal recurrent or folding architectures. Although this problem has not yet been answered in the feed-forward case, efficient algorithms for the training of fixed sigmoidal feed-forward architectures exist. Therefore, this problem is expected to be of equal difficulty as training feed-forward perceptron networks. On the contrary, training recurrent sigmoidal architectures poses several problems in practice and some theoretical investigation of these problems also exist, as we have already mentioned. Additionally, the fundamental behavior of recurrent networks with a sigmoidal or perceptron activation, respectively, is entirely different, as we have seen in the chapter dealing with approximation capabilities. Therefore it is interesting to know, whether this difference leads to a different complexity level of training, too.

Although folding networks are an interesting and well performing approach, a lot of work has to be done before learning machines will be able to solve such complex tasks as football playing or understanding spoken language. For an efficient use of subsymbolic learning methods in all domains, the structure of networks needs further modifications. If we intend to integrate symbolic learning methods and neural networks, we need networks that can produce symbolic data as outputs instead of simple vectors as well. The dual dynamics proposed in the LRAAM unfortunately needs increasing resources even for purely symbolic data. Additionally, more complex data structures like arbitrary graphs occur frequently in applications. Folding networks are usually not adapted to these tasks. For this it is necessary to design different and more complex architectures and training methods and establish the basic theoretical properties of these approaches, too.

Bibliography

1. P. Alexandroff and H. Hopf. *Topologie*, volume 1. Springer, 1974.
2. N. Alon, S. Ben-David, N. Cesa-Bianchi, and D. Haussler. Scale-sensitive dimensions, uniform convergence, and learnability. In *Proceedings of the 34th IEEE Symposium on Foundations of Computer Science*, pp. 292-301, 1993.
3. M. Anthony and J. Shawe-Taylor. A sufficient condition for polynomial distribution-dependent learnability. *Discrete Applied Mathematics*, 77:1-12, 1997.
4. M. Anthony. Uniform convergence and learnability. Technical report, London School of Economics, 1991.
5. M. Anthony. Probabilistic analysis of learning in artificial neural networks: The PAC model and its variants. *Neural Computing Surveys*, 1:1-47, 1997.
6. M. Anthony and N. Biggs. *Computational Learning Theory*. Cambridge Tracts in Theoretical Computer Science. Cambridge University Press, 1992.
7. P. L. Bartlett, P. Long, and R. Williamson. Fat-shattering and the learnability of real valued functions. In *Proceedings of the 7th ACM Conference on Computational Learning Theory*, pp. 299-310, 1994.
8. P. L. Bartlett, V. Maiorov, and R. Meir. Almost linear VC dimension bounds for piecewise polynomial networks. *Neural Computation*, 10(8):2159-2173, 1998.
9. P. Bartlett and R. Williamson. The VC dimension and pseudodimension of two-layer neural networks with discrete inputs. *Neural Computation*, 8(3):653-656, 1996.
10. P. L. Bartlett. For valid generalization, the size of the weights is more important than the size of the network. In M. C. Mozer, M. I. Jordan, and T. Petsche, editors, *Advances in Neural Information Processing Systems*, Volume 9. The MIT Press, pp. 134-141, 1996.
11. S. Basu, R. Pollack, and M.-F. Roy. A new algorithm to find a point in every cell defined by a family of polynomials. *Journal of the ACM*, 43:1002-1045, 1996.
12. E. B. Baum and D. Haussler. What size net gives valid generalization? *Neural Computation*, 1(1):151-165, 1989.
13. S. Ben-David, N. Cesa-Bianci, D. Haussler, and P. Long. Characterizations of learnability for classes of $\{0, \ldots, n\}$-valued functions. *Journal of Computer and System Sciences*, 50:74-86, 1995.
14. Y. Bengio and P. Frasconi. Credit assignment through time: Alternatives to backpropagation. In J. Cowan, G. Tesauro, and J. Alspector, editors, *Advances in Neural Information Processing Systems*, volume 6. Morgan Kaufmann, pp. 75-82, 1994.
15. Y. Bengio and F. Gingras. Recurrent neural networks for missing or asynchronous data. In M. Mozer, D. Touretzky, and M. Perrone, editors, *Advances in Neural Information Processing Systems*, volume 8. The MIT Press, pp. 395-401, 1996.

16. Y. Bengio, P. Simard, and P. Frasconi. Learning long-term dependencies with gradient descent is difficult. *IEEE Transactions on Neural Networks*, 5(2):157-166, 1994.

17. M. Bianchini, S. Fanelli, M. Gori, and M. Maggini. Terminal attractor algorithms: A critical analysis. *Neurocomputing*, 15(1):3-13, 1997.

18. C. Bishop. *Neural Networks for Pattern Recognition*. Oxford University Press, 1995.

19. A. Blum and R. Rivest. Training a 3-node neural network is NP-complete. *Neural Networks*, 9:1017-1023, 1988.

20. H. Braun. *Neuronale Netze*. Springer, 1997.

21. W. L. Buntine and A. S. Weigend. Bayesian back-propagation. *Complex Systems*, 5:603-643, 1991.

22. M. Casey. The dynamics of discrete-time computation, with application to recurrent neural networks and finite state machine extraction. *Neural Computation*, 8(6):1135-1178, 1996.

23. C. Cortes and V. Vapnik. Support vector network. *Machine Learning*, 20:1-20, 1995.

24. F. Costa, P. Frasconi, and G. Soda. A topological transformation for hidden recursive models In M. Verleysen, editor, *European Symposium on Artificial Neural Networks*. D-facto publications, pp. 51-56, 1999.

25. I. Croall and J. Mason. *Industrial Applications of Neural Networks*. Springer, 1992.

26. B. DasGupta, H. T. Siegelmann, and E. D. Sontag. On the complexity of training neural networks with continuous activation. *IEEE Transactions on Neural Networks*, 6(6):1490-1504, 1995.

27. B. Dasgupta and E. D. Sontag. Sample complexity for learning recurrent perceptron mappings. *IEEE Transactions on Information Theory*, 42:1479-1487, 1996.

28. A. Eliseeff and H. Paugam-Moisy. Size of multilayer networks for exact learning. In M. C. Mozer, M. I. Jordan, and T. Petsche, editors, *Advances in Neural Information Processing Systems*, volume 9. The MIT Press, pp. 162-168, 1996.

29. J. L. Elman. Finding structure in time. *Cognitive Science*, 14:179-211, 1990.

30. J. L. Elman. Distributed representations, simple recurrent networks, and grammatical structure. *Machine Learning*, 7:195-225, 1991.

31. T. Elsken. Personal communication.

32. S. E. Fahlman. An empirical study of learning speed in back-propagation networks. In *Proceedings of the 1988 Connectionist Models Summer School*. Morgan Kaufmann, 1988

33. S. E. Fahlman. The recurrent cascade-correlation architecture. In R. Lippmann, J. Moody, D. Touretzky, and S. Hanson, editors, *Advances in Neural Information Processing Systems*, volume 3. Morgan Kaufmann, pp. 190-198, 1991.

34. J. Fodor and Z. Pylyshin. Connectionism and cognitive architecture: A critical analysis. *Cognition*, 28:3-71, 1988.

35. P. Frasconi, M. Gori, S. Fanelli, and M. Protasi. Suspiciousness of loading problems. In *IEEE International Conference on Neural Networks*, 1997.

36. P. Frasconi, M. Gori, and A. Sperduti. A general framework for adaptive processing of data sequences. *IEEE Transactions on Neural Networks*, 9(5):768-786, 1997.

37. M. Garey and D. Johnson. *Computers and Intractability: A Guide to the Theory of NP-Completeness*. W. H. Freeman and Company, 1979.

38. C. L. Giles and M. Gori, editors. *Adaptive Processing of Sequences and Data Structures*. Springer, 1998.

39. C. L. Giles, G. M. Kuhn, and R. J. Williams. Special issue on dynamic recurrent neural networks. *IEEE Transactions on Neural Networks*, 5(2), 1994.

40. C. L. Giles and C. W. Omlin. Pruning recurrent neural networks for improved generalization performance. *IEEE Transactions on Neural Networks*, 5(5):848-851, 1994.

41. C. Goller. *A connectionist approach for learning search control heuristics for automated deduction systems.* PhD thesis, Technische Universität München, 1997.

42. C. Goller and A. Küchler. Learning task-dependent distributed representations by backpropagation through structure. In *Proceedings of the IEEE Conference on Neural Networks*, pp. 347-352, 1996.

43. M. Gori, M. Mozer, A. C. Tsoi, and R. L. Watrous. Special issue on recurrent neural networks for sequence processing. *Neurocomputing*, 15(3-4), 1997.

44. L. Gurvits and P. Koiran. Approximation and learning of convex superpositions. In *2nd European Conference on Computational Learning Theory*, pp. 222-236, 1995.

45. B. Hammer. On the learnability of recursive data. *Mathematics of Control, Signals, and Systems*, 12:62-79, 1999.

46. B. Hammer. On the generalization of Elman networks. In W. Gerstner, A. Germond, M. Hasler, and J.-D. Nicaud, editors, *Artificial Neural Networks – ICANN'97*. Springer, pp. 409-414, 1997.

47. B. Hammer. On the approximation capability of recurrent neural networks. In M. Heiss, editor, *Proceedings of the International Symposium on Neural Computation*. ICSC Academic Press, pp. 512-518, 1998.

48. B. Hammer. Some complexity results for perceptron networks. In L. Niklasson, M. Bodén, and T. Ziemke, editors, *Proceedings of the 8th International Conference on Artificial Neural Networks*. Springer, pp. 639-644, 1998.

49. B. Hammer. Training a sigmoidal network is difficult. In M. Verleysen, editor, *European Symposium on Artificial Neural Networks*. D-facto publications, pp. 255-260, 1998.

50. B. Hammer. Approximation capabilities of folding networks. In M. Verleysen, editor, *European Symposium on Artificial Neural Networks*. D-facto publications, pp. 33-38, 1999.

51. D. Haussler. Decision theoretic generalizations of the PAC model for neural net and other learning applications. *Information and Computation*, 100:78-150, 1992.

52. D. Haussler, M. Kearns, N. Littlestone, and M. Warmuth. Equivalence of models for polynomial learnability. *Information and Computation*, 95:129-161, 1991.

53. J. Hertz, A. Krogh, and R. Palmer. *Introduction to the Theory of Neural Computation*. Addison Wesley, 1991.

54. S. Hochreiter and J. Schmidhuber. Long short-term memory. *Neural Computation*, 9(8):1735-1780, 1997.

55. K.-U. Höffgen. Computational limitations on training sigmoidal neural networks. *Information Processing Letters*, 46(6):269-274, 1993.

56. K.-U. Höffgen, H.-U. Simon, and K. S. VanHorn. Robust trainability of single neurons. *Journal of Computer and System Sciences*, 50:114-125, 1995.

57. B. G. Horne and C. L. Giles. An experimental comparison of recurrent neural networks. In G. Tesauro, D. Touretzky, and T. Leen, editors, *Advances in Neural Information Processing Systems*, volume 7. The MIT Press, pp. 697-704, 1995.

58. K. Hornik. Some new results on neural network approximation. *Neural Networks*, 6:1069-1072, 1993.

59. K. Hornik, M. Stinchcombe, and H. White. Multilayer feedforward networks are universal approximators. *Neural Networks*, 2:359-366, 1989.

60. B. Jaehne. *Digitale Bildverarbeitung*. Springer, 1989.

61. L. Jones. The computational intractability of training sigmoidal neural networks. *IEEE Transactions on Information Theory*, 43(1):167-173, 1997.

62. S. Judd. *Neural Network Design and the Complexity of Learning*. MIT Press, 1990.

63. N. Karmarkar. A new polynomial time algorithm for linear programming. *Combinatorica*, 4(4):373-395, 1984.

64. M. Karpinski and A. Macintyre. Polynomial bounds for the VC dimension of sigmoidal neural networks. In *Proceedings of the 27th annual ACM Symposium on the Theory of Computing*, pp. 200-208, 1995.

65. L. Khachiyan. A polynomial algorithm for linear programming. *Soviet Mathematics Doklady*, 20:191-194, 1979.

66. J. Kilian and H. T. Siegelmann. The dynamic universality of sigmoidal neural networks. *Information and Computation*, 128:48-56, 1996.

67. L. Kindermann. An addition to backpropagation for computing functional roots. In M. Heiss, editor, *International Symposium on Neural Computation*. ICSC Academic Press, pp.424-427, 1998.

68. P. Koiran, M. Cosnard, and M. Garzon. Computability with low-dimensional dynamical systems. *Theoretical Computer Science*, 132:113-128, 1994.

69. P. Koiran and E. D. Sontag. Neural networks with quadratic VC dimension. *Journal of Computer and System Sciences*, 54:223-237, 1997.

70. P. Koiran and E. D. Sontag. Vapnik-Chervonenkis dimension of recurrent neural networks. In *Proceedings of the 3rd European Conference on Computational Learning Theory*, pp. 223-237, 1997.

71. C.-M. Kuan, K. Hornik, and H. White. A convergence result in recurrent neural networks. *Neural Computation*, 6(3):420-440, 1994.

72. A. Küchler. On the correspondence between neural folding architectures and tree automata. Technical report, University of Ulm, 1998.

73. A. Küchler and C. Goller. Inductive learning symbolic domains using structure-driven neural networks. In G. Görz and S. Hölldobler, editors, *KI-96: Advances in Artificial Intelligence*. Springer, pp. 183-197, 1996.

74. S. R. Kulkarni, S. K. Mitter, and J. N. Tsitsiklis. Active learning using arbitrary binary queries. *Machine Learning*, 11:23-35, 1993.

75. S. R. Kulkarni and M. Vidyasagar. Decision rules for pattern classification under a family of probability measures. *IEEE Transactions on Information Theory*, 43:154-166, 1997.

76. M. C. Laskowski. Vapnik-Chervonenkis classes of definable sets. *Journal of the London Mathematical Society*, 45:377-384, 1992.

77. Y. LeCun, J. Denker, and S. Solla. Optimal brain damage. In D. Touretzky, editor, *Advances in Neural Information Processing Systems*, volume 2. Morgan Kaufmann, pp. 598-605, 1990.

78. J.-H. Lin and J. Vitter. Complexity results on learning by neural networks. *Machine Learning*, 6:211-230, 1991.

79. T. Lin, B. Horne, P. Tino, and C. L. Giles. Learning long-term dependencies is not as difficult with NARX recurrent neural networks. Technical report, University of Maryland, 1995.

80. R. Lippmann. Review of neural networks for speech recognition. *Neural Computation*, 1:1-38, 1989.

81. D. W. Loveland. Mechanical theorem-proving for model elimination. *Journal of the ACM* 15(2):236-251, 1968.

82. W. Maass. Neural nets with superlinear VC-dimension. *Neural Computation*, 6:877-884, 1994.

83. W. Maass. Agnostic PAC-learning of functions on analog neural nets. *Neural Computation*, 7(5):1054-1078, 1995.

84. W. Maass and P. Orponen. On the effect of analog noise in discrete-time analog computation. *Neural Computation*, 10(5):1071-1095, 1998.

85. W. Maass and E. Sontag. Analog neural nets with Gaussian or other common noise distributions cannot recognize arbitrary regular languages. *Neural Computation*, 11:771-782, 1999.

86. A. Macintyre and E. D. Sontag. Finiteness results for sigmoidal 'neural' networks. In *Proceedings 25th Annual Symposium Theory Computing*, pp. 325-334, 1993.

87. M. Masters. *Neural, Novel, & Hybrid Algorithms for Time Series Prediction*. Wiley, 1995.

88. U. Matecki. *Automatische Merkmalsauswahl für Neuronale Netze mit Anwendung in der pixelbezogenen Klassifikation von Bildern*. PhD thesis, University of Osnabrück, 1999.

89. N. Megiddo. On the complexity of polyhedral separability. *Discrete and Computational Geometry*, 3:325-337, 1988.

90. M. Minsky and S. Papert. *Perceptrons*. MIT Press, 1988.

91. M. Mozer. Neural net architectures for temporal sequence processing. In A. Weigend and N. Gershenfeld, editors, *Predicting the future and understanding the past*. Addison-Wesley, pp. 143-164, 1993.

92. M. Mozer and P. Smolensky. Skeletonization: A technique for trimming the fat from a network via relevant assessment. In D. Touretzky, editor, *Advances in Neural Information Processing Systems*, volume 1. Morgan Kaufmann, pp. 107-115, 1989.

93. K. S. Narendra and Parthasarathy. Identification and control of dynamical networks. *IEEE Transactions on Neural Networks*, 1(1):4-27, 1990.

94. B. K. Natarajan. *Machine learning: A theoretical approach*. Morgan Kaufmann, 1991.

95. A. Nobel and A. Dembo. A note on uniform laws of averages for dependent processes. *Statistics and Probability Letters*, 17:169-172, 1993.

96. S. J. Nowlan and G. E. Hinton. Simplifying neural networks by soft weight sharing. *Neural Computation*, 4(4):473-493, 1992.

97. C. Omlin and C. Giles. Constructing deterministic finite-state automata in recurrent neural networks. *Journal of the ACM*, 43(2):937-972, 1996.

98. B. A. Pearlmutter. Gradient calculations for dynamic recurrent neural networks: A survey. *IEEE Transactions on Neural Networks*, 6(5):1212-1228, 1995.

99. L. Pitt and L. Valiant. Computational limitations on learning from examples. *Journal of the Association for Computing Machinery*, 35(4):965-984, 1988.

100. T. Plate. Holographic reduced representations. *IEEE Transactions on Neural Networks*, 6(3):623-641, 1995.

101. J. Pollack. Recursive distributed representation. *Artificial Intelligence*, 46(1-2):77-106, 1990.

102. M. Reczko. Protein secondary structure prediction with partially recurrent neural networks. *SAR and QSAR in environmental research*, 1:153-159, 1993.

103. M. Riedmiller and H. Braun. A direct adaptive method for faster backpropagation: The RPROP algorithm. In *Proceedings of the Sixth International Conference on Neural Networks*. IEEE, pp. 586-591, 1993.

104. J. A. Robinson. A machine oriented logic based on the resolution principle. *Journal of the ACM*, 12(1):23-41, 1965.

105. D. Rumelhart, G. Hinton, and R. Williams. Learning representations by back-propagating errors. *Nature*, 323:533-536, 1986.
106. D. Rumelhart, G. Hinton, and R. Williams. Learning internal representations by back-propagating errors. In *Neurocomputing: Foundations of Research*. MIT Press, pp. 696-700, 1988.
107. J. Schmidhuber. A fixed size storage $O(m^3)$ time complexity learning algorithm for fully recurrent continually running networks. *Neural Computation*, 4(2):243-248, 1992.
108. M. Schmitt. *Komplexität neuronaler Lernprobleme*. Peter Lang, 1996.
109. M. Schmitt. Proving hardness results of neural network training problems. *Neural Networks*, 10(8):1533-1534, 1997.
110. T. Schmitt and C. Goller. Relating chemical structure to activity with the structure processing neural folding architecture. In *Engineering Applications of Neural Networks*, 1998.
111. S. Schulz, A. Küchler, and C. Goller. Some experiments on the applicability of folding architectures to guide theorem proving. In *Proceedings of the 10th International FLAIRS Conference*, pp. 377-381, 1997.
112. T. Sejnowski and C. Rosenberg. Parallel networks that learn to pronounce English text. *Complex Systems*, 1:145-168, 1987.
113. J. Shawe-Taylor, P. L. Bartlett, R. Williamson, and M. Anthony. Structural risk minimization over data dependent hierarchies. Technical report, Neuro-COLT, 1996.
114. H. T. Siegelmann. The simple dynamics of super Turing theories. *Theoretical Computer Science*, 168:461-472, 1996.
115. H. T. Siegelmann and E. D. Sontag. Analog computation, neural networks, and circuits. *Theoretical Computer Science*, 131:331-360, 1994.
116. H. T. Siegelmann and E. D. Sontag. On the computational power of neural networks. *Journal of Computer and System Sciences*, 50:132-150, 1995.
117. P. Smolensky. Tensor product variable binding and the representation of symbolic structures in connectionist systems. *Artificial Intelligence*, 46(1-2):159-216, 1990.
118. E. D. Sontag. VC dimension of neural networks. In C. Bishop, editor, *Neural Networks and Machine Learning*. Springer, pp. 69-95, 1998.
119. E. D. Sontag. Feedforward nets for interpolation and classification. *Journal of Computer and System Sciences*, 45:20-48, 1992.
120. E. D. Sontag. Neural nets as systems models and controllers. In *7th Yale Workshop on Adaptive and Learning Systems*, pp. 73-79, 1992.
121. F. Soulie and P. Gallinari. *Industrial Applications of Neural Networks*. World Scientific, 1998.
122. A. Sperduti. Labeling RAAM. *Connection Science*, 6(4):429-459, 1994.
123. A. Sperduti. On the computational power of recurrent neural networks for structures. *Neural Networks*, 10(3):395, 1997.
124. A. Sperduti and A. Starita. Dynamical neural networks construction for processing of labeled structures. Technical report, University of Pisa, 1995.
125. M. Stone. Cross-validatory choice and assessment of statistical predictions (with discussion). *Journal of the Royal Statistical Society B*, 36:111-147, 1974.
126. J. Suykens, B. DeMoor, and J. Vandewalle. Static and dynamic stabilizing neural controllers applicable to transition between equilibrium point. *Neural Networks*, 7(5):819-831, 1994.
127. P. Tino, B. Horne, C. L. Giles, and P. Colligwood. Finite state machines and recurrent neural networks – automata and dynamical systems approaches. In *Neural Networks and Pattern Recognition*. Academic Press, pp. 171-220, 1998.

128. D. Touretzky. BoltzCONS: Dynamic symbol structures in a connectionist network. *Artificial Intelligence*, 46:5-46, 1990.

129. L. Valiant. A theory of the learnable. *Communications of the ACM*, 27(11):1134-1142, 1984.

130. V. Vapnik. *The Nature of Statistical Learning Theory*. Springer, 1995.

131. V. Vapnik and A. Chervonenkis. On the uniform convergence of relative frequencies of events to their probabilities. *Theory of Probability and its Applications*, 16(2):264-280, 1971.

132. M. Vidyasagar. *A Theory of Learning and Generalization*. Springer, 1997.

133. M. Vidyasagar. An introduction to the statistical aspects of PAC learning theory. *Systems and Control Letters*, 34:115-124, 1998.

134. J. Šíma. Back-propagation is not efficient. *Neural Networks*, 9(6):1017-1023, 1996.

135. V. H. Vu. On the infeasibility of training neural networks with small squared error. In M. C. Mozer, M. I. Jordan, and T. Petsche, editors, *Advances in Neural Information Processing Systems*, volume 10. The MIT Press, pp. 371-377, 1998.

136. P. Werbos. *Beyond Regression: New Tools for Prediction and Analysis in the Behavioral Science*. PhD thesis, Harvard University, 1974.

137. P. Werbos. Backpropagation through time: What it does and how to do it. *Proceedings of the IEEE*, 78(10):1550-1560, 1990.

138. P. Werbos. *The roots of backpropagation*. Wiley, 1994.

139. R. Williams and D. Zipser. Gradient-based learning algorithms for recurrent networks and their computational complexity. In Y. Chauvin and D. Rumelhart, editors, *Back-propagation: Theory, Architectures and Applications*. Erlbaum, pp. 433-486, 1995.

140. W. H. Wilson. A comparison of architectural alternatives for recurrent networks. In *Proceedings of the Fourth Australian Conference on Neural Networks*, pp. 189-192, 1993.

141. A. Zell. *Simulation Neuronaler Netze*. Addison-Wesley, 1994.

Index

Lecture Notes in Control and Information Sciences

Edited by M. Thoma

1997–2000 Published Titles:

Vol. 240: Lin, Z.
Low Gain Feedback
376 pp. 1999 [1-85233-081-3]

Vol. 241: Yamamoto, Y.; Hara S.
Learning, Control and Hybrid Systems
472 pp. 1999 [1-85233-076-7]

Vol. 242: Conte, G.; Moog, C.H.; Perdon A.M.
Nonlinear Control Systems
192 pp. 1999 [1-85233-151-8]

Vol. 243: Tzafestas, S.G.; Schmidt, G. (Eds)
Progress in Systems and Robot Analysis and Control Design
624 pp. 1999 [1-85233-123-2]

Vol. 244: Nijmeijer, H.; Fossen, T.I. (Eds)
New Directions in Nonlinear Observer Design
552pp: 1999 [1-85233-134-8]

Vol. 245: Garulli, A.; Tesi, A.; Vicino, A. (Eds)
Robustness in Identification and Control
448pp: 1999 [1-85233-179-8]

Vol. 246: Aeyels, D.;
Lamnabhi-Lagarrigue, F.; van der Schaft, A. (Eds)
Stability and Stabilization of Nonlinear Systems
408pp: 1999 [1-85233-638-2]

Vol. 247: Young, K.D.; Özgüner, Ü. (Eds)
Variable Structure Systems, Sliding Mode and Nonlinear Control
400pp: 1999 [1-85233-197-6]

Vol. 248: Chen, Y.; Wen C.
Iterative Learning Control
216pp: 1999 [1-85233-190-9]

Vol. 249: Cooperman, G.; Jessen, E.; Michler, G. (Eds)
Workshop on Wide Area Networks and High Performance Computing
352pp: 1999 [1-85233-642-0]

Vol. 250: Corke, P. ; Trevelyan, J. (Eds)
Experimental Robotics VI
552pp: 2000 [1-85233-210-7]

Vol. 251: van der Schaft, A. ; Schumacher, J.
An Introduction to Hybrid Dynamical Systems
192pp: 2000 [1-85233-233-6]

Vol. 252: Salapaka, M.V.; Dahleh, M.
Multiple Objective Control Synthesis
192pp. 2000 [1-85233-256-5]

Vol. 253: Elzer, P.F.; Kluwe, R.H.; Boussoffara, B.
Human Error and System Design and Management
240pp. 2000 [1-85233-234-4]